ASE Test Preparation

Medium/Heavy Duty Truck Technician Certification Series

Heating, Ventilation & Air Conditioning (T7)
5th Edition

DELMAR
CENGAGE Learning·

Australia • Brazil • Japan • Korea • Mexico • Singapore • Spain • United Kingdom • United States

DELMAR
CENGAGE Learning

ASE Test Preparation: Medium/Heavy Duty Truck Technician Certification Series, Heating, Ventilation & Air Conditioning (T7), 5ᵗʰ Edition

Vice President, Technology and Trades Professional Business Unit: Gregory L. Clayton

Director, Professional Transportation Industry Training Solutions: Kristen L. Davis

Editorial Assistant: Danielle Filippone

Director of Marketing: Beth A. Lutz

Marketing Manager: Jennifer Barbic

Senior Production Director: Wendy Troeger

Production Manager: Sherondra Thedford

Content Project Management: PreMediaGlobal

Senior Art Director: Benjamin Gleeksman

Section Opener Image: Evgeny Korshenkov, 2012. Used under license from Shutterstock.com

ISBN-13: 978-1-111-12903-3

ISBN-10: 1-111-12903-7

Delmar Cengage Learning
5 Maxwell Drive
Clifton Park, NY 12065-2919
USA

Cengage Learning is a leading provider of customized learning solutions with office locations around the globe, including Singapore, the United Kingdom, Australia, Mexico, Brazil, and Japan. Locate your local office at: **international.cengage.com/region**.

Cengage Learning products are represented in Canada by Nelson Education, Ltd.

For more information on transportation titles available from Delmar, Cengage Learning, please visit our website at **www.trainingbay.cengage.com**.

For more learning solutions, please visit our corporate website at **www.cengage.com**.

Notice to the Reader

Printed in the United States of America
1 2 3 4 5 6 7 15 14 13 12 11

Table of Contents

Delmar, a part of Cengage Learning, is very pleased that you have chosen to use our ASE Test Preparation Guide to help prepare yourself for the Heating, Ventilation & Air Conditioning (T7) ASE certification examination. This guide is designed to help prepare you for your actual exam by providing you with an overview and introduction of the testing process, introducing you to the task list for the Heating, Ventilation & Air Conditioning (T7) certification exam, giving you an understanding of what knowledge and skills you are expected to have in order to successfully perform the duties associated with each task area, and providing you with several preparation exams designed to emulate the live exam content in hopes of assessing your overall exam readiness.

If you have a basic working knowledge of the discipline you are testing for, you will find this book is an excellent guide, helping you understand the "must know" items needed to successfully pass the ASE certification exam. This manual is not a textbook. Its objective is to prepare the individual who has the existing requisite experience and knowledge to attempt the challenge of the ASE certification process. This guide cannot replace the hands-on experience and theoretical knowledge required by ASE to master the vehicle repair technology associated with this exam. If you are unable to understand more than a few of the preparation questions and their corresponding explanations in this book, it could be that you require either more shop-floor experience or further study.

This book begins by providing an overview of, and introduction to, the testing process. This section outlines what we recommend you do to prepare, what to expect on the actual test day, and overall methodologies for your success. This section is followed by a detailed overview of the ASE task list to include explanations of the knowledge and skills you must possess to successfully answer questions related to each particular task. After the task list, we provide six sample preparation exams for you to use as a means of evaluating areas of understanding, as well as areas requiring improvement in order to successfully pass the ASE exam. Delmar is the first and only test preparation organization to provide so many unique preparation exams. We enhanced our guides to include this support as a means of providing you with the best preparation product available. Section 6 of this guide includes the answer keys for each preparation exam, along with the answer explanations for each question. Each answer explanation also contains a reference back to the related task or tasks that it assesses. This will provide you with a quick and easy method for referring back to the task list whenever needed. The last section of this book contains blank answer sheet forms you can use as you attempt each preparation exam, along with a glossary of terms.

OUR COMMITMENT TO EXCELLENCE

Thank you for choosing Delmar, Cengage Learning for your ASE test preparation needs. All of the writers, editors, and Delmar staff have worked very hard to make this test preparation guide second to none. We feel confident that you will find this guide easy to use and extremely beneficial as you prepare for your actual ASE exam.

Delmar, Cengage Learning has sought out the best subject matter experts in the country to help with the development of *ASE Test Preparation: Medium/Heavy Duty Truck Technician Certification Series, Heating, Ventilation & Air Conditioning (T7), 5th Edition*. Preparation questions are authored and then reviewed by a group of certified, subject-matter experts to ensure the highest level of quality and validity to our product.

If you have any questions concerning this guide or any guide in this series, please visit us on the web at **http://www.trainingbay.cengage.com**.

For web-based online test preparation for ASE certifications; please visit us on the web at **http://www.techniciantestprep.com/** to learn more.

ABOUT THE AUTHOR

Jerry Clemons has been around cars, trucks, equipment, and machinery throughout his whole life. Being raised on a large farm in central Kentucky provided him with an opportunity to complete mechanical repair procedures from an early age. Jerry earned an associate in applied science degree in Automotive Technology from Southern Illinois University and a bachelor of science degree in Vocational, Industrial, and Technical Education from Western Kentucky University. Jerry has also completed a master of science degree in Safety, Security, and Emergency Management from Eastern Kentucky University. Jerry has been employed at Elizabethtown Community and Technical College since 1999 and is currently an associate professor for the Automotive and Diesel Technology Programs. Jerry holds the following ASE certifications: Master Medium/Heavy Truck Technician, Master Automotive Technician, Advanced Engine Performance (L1), Truck Equipment Electrical Installation (E2), and Automotive Service Consultant (C1). Jerry is a member of the Mobile Air Conditioning Society (MACS) as well as a member of the North American Council of Automotive Teachers (NACAT). Jerry has been involved in developing transportation material for Cengage Learning for seven years.

ABOUT THE SERIES ADVISOR

Brian (BJ) Crowley has experienced several different aspects of the diesel industry over the past 10 years. Now a diesel technician in the oil and gas industry, BJ owned and operated a diesel repair shop where he repaired heavy, medium, and light trucks, as well as agricultural and construction equipment. He earned an associate's degree in diesel technology from Elizabethtown Community and Technical College in Kentucky and is an ASE Master certified medium/heavy duty truck technician.

The History and Purpose of ASE

ASE began as the National Institute for Automotive Service Excellence (NIASE). It was founded as a non-profit, independent entity in 1972 by a group of industry leaders with the single goal of providing a means for consumers to distinguish between incompetent and competent technicians. It accomplishes this goal through the testing and certification of repair and service professionals. Though it is still known as the National Institute for Automotive Service Excellence, it is now called "ASE" for short.

Today, ASE offers more than 40 certification exams in automotive, medium/heavy duty truck, collision repair and refinish, school bus, transit bus, parts specialist, automobile service consultant, and other industry-related areas. At this time there are more than 385,000 professionals nationwide with current ASE certifications. These professionals are employed by new car and truck dealerships, independent repair facilities, fleets, service stations, franchised service facilities, and more.

ASE's certification exams are industry-driven and cover practically every on-highway vehicle service segment. The exams are designed to stress the knowledge of job-related skills. Certification consists of passing at least one exam and documenting two years of relevant work experience. To maintain certification, those with ASE credentials must be re-tested every five years.

While ASE certifications are a targeted means of acknowledging the skills and abilities of an individual technician, ASE also has a program designed to provide recognition for highly qualified repair, support, and parts businesses. The Blue Seal of Excellence Recognition Program allows businesses to showcase their technicians and their commitment to excellence. One of the requirements of becoming Blue Seal recognized is that the facility must have a minimum of 75 percent of its technicians ASE certified. Additional criteria apply, and program details can be found on the ASE website.

ASE recognized that educational programs serving the service and repair industry also needed a way to be recognized as having the faculty, facilities, and equipment to provide a quality education to students wanting to become service professionals. Through the combined efforts of ASE, industry, and education leaders, the non-profit organization entitled the National Automotive Technicians Education Foundation (NATEF) was created in 1983 to evaluate and recognize academic programs. Today more than 2,000 educational programs are NATEF certified.

For additional information about ASE, NATEF, or any of their programs, the following contact information can be used:

National Institute for Automotive Service Excellence (ASE)

101 Blue Seal Drive S.E.

Suite 101

Leesburg, VA 20175

Telephone: 703-669-6600

Fax: 703-669-6123

Website: **www.ase.com**

Overview and Introduction

Participating in the National Institute for Automotive Service Excellence (ASE) voluntary certification program provides you with the opportunity to demonstrate you are a qualified and skilled technician that has the "know-how" required to successfully work on today's modern vehicles.

EXAM ADMINISTRATION

> *Note:* After November 2011, ASE will no longer offer paper and pencil certification exams. There will be no Winter testing window in 2012, and ASE will offer and support CBT testing exclusively starting in April 2012.

ASE provides computer-based testing (CBT) exams, which are administered at test centers across the nation. It is recommended that you go to the ASE website at *http://www.ase.com* and review the conditions and requirements for this type of exam. There is also an exam demonstration page that allows you to personally experience how this type of exam operates before you register.

CBT exams are available four times annually, for two-month windows, with a month of no testing in between each testing window:

- January/February – Winter testing window
- April/May – Spring testing window
- July/August – Summer testing window
- October/November – Fall testing window

Please note, testing windows and timing may change. It is recommended you go to the ASE website at *http://www.ase.com* and review the latest testing schedules.

UNDERSTANDING TEST QUESTION BASICS

ASE exam questions are written by service industry experts. Each question on an exam is created during an ASE-hosted "item-writing" workshop. During these workshops, expert service representatives from manufacturers (domestic and import), aftermarket parts and equipment manufacturers, working technicians, and vocational educators gather to share ideas and convert them into actual exam questions. Each exam question written by these experts must then survive review by all members of the group. The questions are designed to address the practical application of repair and diagnosis knowledge and skills practiced by technicians in their day-to-day work.

After the item-writing workshop, all questions are pre-tested and quality-checked on a national sample of technicians. Those questions that meet ASE standards of quality and accuracy are included in the scored sections of the exams; the "rejects" are sent back to the drawing board or discarded altogether.

Depending on the topic of the certification exam, you will be asked between 40 and 80 multiple-choice questions. You can determine the approximate number of questions you can expect to be asked during the Heating, Ventilation & Air Conditioning (T7) certification exam by reviewing the task list in Section 4 of this book. The five-year recertification exam will cover this same content; however, the number of questions for each content area of the recertification exam will be reduced by approximately one-half.

> *Note:* Exams may contain questions that are included for statistical research purposes only. Your answers to these questions will not affect your score, but since you do not know which ones they are, you should answer all questions in the exam.

Using multiple criteria, including cross-sections by age, race, and other background information, ASE is able to guarantee that exam questions do not include bias for or against any particular group. A question that shows bias toward any particular group is discarded.

TEST-TAKING STRATEGIES

Before beginning your exam, quickly look over the exam to determine the total number of questions that you will need to answer. Having this knowledge will help you manage your time throughout the exam to ensure you have enough time available to answer all of the questions presented. Read through each question completely before marking your answer. Answer the questions in the order they appear on the exam. Leave the questions blank that you are not sure of and move on to the next question. You can return to those unanswered questions after you have finished the others. These questions may actually be easier to answer at a later time, once your mind has had additional time to consider them on a subconscious level. In addition, you might find information in other questions that will help you recall the answers to some of them.

Multiple-choice exams are sometimes challenging because there are often several choices that may seem possible, or partially correct, and therefore it may be difficult to decide on the most appropriate answer choice. The best strategy, in this case, is to first determine the correct answer before looking at the answer options. If you see the answer you decided on, you should still be careful to examine the other answer options to make sure that none seem more correct than yours. If you do not know or are not sure of the answer, read each option very carefully and try to eliminate those options that you know are incorrect. That way, you can often arrive at the correct choice through a process of elimination.

If you have gone through the entire exam and you still do not know the answer to some of the questions, *then guess.* Yes, guess. You then have at least a 25 percent chance of being correct. While your score is based on the number of questions answered correctly, any question left blank, or unanswered, is automatically scored as incorrect.

There is a lot of "folk" wisdom on the subject of test taking that you may hear about as you prepare for your ASE exam. For example, there are those who would advise you to avoid response options that use certain words such as *all, none, always, never, must,* and *only,* to name a few. This, they claim, is because nothing in life is exclusive. They would advise you to choose response options that use words that allow for some exception, such as *sometimes, frequently, rarely, often, usually, seldom,* and *normally.* They would also advise you to avoid the first and last option (A or D) because exam writers, they feel, are more comfortable if they put the correct answer in the middle (B or C) of the choices. Another recommendation often offered is to select the option that is either shorter or longer than the other three choices because it is more likely to be correct. Some would advise you to never change an answer since your first intuition is usually correct. Another area of "folk" wisdom focuses specifically on any repetitive patterns created by your question responses (i.e., A, B, C, A, B, C, A, B, C).

Many individuals may say that there are actual grains of truth in this "folk" wisdom, and whereas with some exams, this may prove true, it is not relevant with regard to the ASE certification exams. ASE validates all exam questions and test forms through a national sample of technicians and only those questions and test forms that meet ASE standards of quality and accuracy are included in the scored sections of the exams. Any biased questions or patterns are discarded altogether and therefore, it is highly unlikely you will actually experience any of this "folk" wisdom on an actual ASE exam.

PREPARING FOR THE EXAM

Delmar, Cengage Learning wants to make sure we are providing you with the most thorough preparation guide possible. To demonstrate this, we have included hundreds of preparation questions in this guide. These questions are designed to provide as many opportunities as possible to prepare you to successfully attempt and pass your ASE exam. The preparation approach we recommend and outline in this book is designed to help you build confidence in demonstrating what task area content you already know well, while also outlining what areas you should review in more detail prior to the actual exam.

We recommend that your first step in the preparation process should be to thoroughly review Section 3 of this book. This section contains a description and explanation of the type of questions you will find on an ASE exam.

Once you understand how the questions will be presented, we then recommend that you thoroughly review Section 4 of this book. This section contains information that will help you establish an understanding of what the exam will be evaluating, and specifically, how many questions to expect to be asked in each specific task area.

As your third preparatory step, we recommend you attempt your first preparation exam, located in Section 5 of this book. Answer one question at a time. After you answer each question, review the answer and question explanation information, located in Section 6. This section will provide you with instant feedback, allowing you to gauge your progress, one question at a time, throughout this first preparation exam attempt. If after reading the question explanation you do not feel you understand the reasoning for the correct answer, go back and review the task list overview (Section 4) for the task that is related to that question. Included with each question explanation is a clear identifier of the task area that is being assessed (e.g., Task A.1). If at that point you still do not feel you have a solid understanding of the material, identify a good source of information on the topic, such as an educational course, textbook, or other related source of topical learning, and do some additional studying.

After you have completed your first preparation exam and have reviewed your answers, you are ready to attempt your next preparation exam. A total of six practice exams are available in Section 5 of this book. For your second preparation exam, we recommend that you answer the questions as if you were taking the actual exam. Do not use any reference material or allow any interruptions in order to get a feel for how you will do on the actual exam. Once you have answered all of the questions, grade your results using the answer key in Section 6. For every question that you gave an incorrect answer to, study the explanations to the answers and/or the overview of the related task areas. Try to determine the root cause for your missing the question. The easiest thing to correct is learning the correct technical content. The hardest things to correct are behaviors that lead you to an incorrect conclusion. If you knew the information but still got the question incorrect, there is likely a test-taking behavior that will need to be corrected. An example of this would be reading too quickly and skipping over words that affect your reasoning. If you can identify what you did that caused you to answer the question incorrectly, you can eliminate that cause and improve your score.

Here are some basic guidelines to follow while preparing for the exam:

- Focus your studies on those areas you are weak in.
- Be honest with yourself when determining if you understand something.
- Study often but for short periods of time.
- Remove yourself from all distractions when studying.
- Keep in mind the goal of studying is not just to pass the exam; the real goal is to learn.
- Prepare physically by getting a good night's rest before the exam and eat meals that provide energy, but do not cause discomfort.
- Arrive early to the exam site to avoid long waits as test candidates check in.
- Use all of the time available for your exams. If you finish early, spend the remaining time reviewing your answers.
- Do not leave any questions unanswered. If absolutely necessary, guess. All unanswered questions are automatically scored as incorrect.

Here are some items you will need to bring to the exam site:

- A valid, government- or school-issued photo ID
- Your test center admissions ticket
- A watch (not all test sites have clocks)

> *Note:* Books, calculators, and other reference materials are not allowed in the exam room. The exceptions to this list are English-Foreign dictionaries or glossaries. All items will be inspected before and after testing.

WHAT TO EXPECT DURING THE EXAM

When taking a CBT exam, as soon as you are seated in the testing center, you will be given a brief tutorial to acquaint you with the computer-delivered test prior to taking your certification exam(s). The CBT exams allow you to only select one answer per question. You can also change your answers as many times as you like. When you select a second answer choice, the CBT will automatically unselect your first answer choice. If you want to skip a question to return to later, you can utilize the "flag" feature, which will allow you to quickly identify and review questions whenever you are ready. Prior to completing your exam, you will also be provided with an opportunity to review your answers and address any unanswered questions.

TESTING TIME

Each individual ASE CBT exam has a fixed time limit. Individual exam times will vary based upon exam area and will range anywhere from a half hour to two hours. You will also be given an additional 30 minutes beyond what is allotted to complete your exams to ensure you have adequate time to perform all necessary check-in procedures, complete a brief CBT tutorial, and potentially complete a post-test survey.

You can register for and take multiple CBT exams during one testing appointment. The maximum time allotment for a CBT appointment is four and a half hours. If you happen to register for so many exams that you will require more time than this, your exams will be scheduled into multiple appointments. This could mean that you have testing on both the morning and the afternoon of the same day, or they could be scheduled on different days, depending on your personal preference and the test center's schedule.

It is important to understand that if you arrive late for your CBT test appointment, you will not be able to make up any missed time. You will only have the scheduled amount of time remaining in your appointment to complete your exam(s).

While most people finish their CBT exams within the time allowed, others might feel rushed or not be able to finish the test, due to the implied stress of a specific, individual time limit allotment. Before you register for the CBT exams, you should review the number of exam questions that will be asked along with the amount of time allotted for that exam to determine whether you feel comfortable with the designated time limitation or not.

As an overall time management recommendation, you should monitor your progress and set a time limit you will follow with regard to how much time you will spend on each individual exam question. This should be based on the total number of questions you will be answering.

Also, it is very important to note that if for any reason you wish to leave the testing room during an exam, you must first ask permission. If you happen to finish your exam(s) early and wish to leave the testing site before your designated session appointment is completed, you are permitted to do so only during specified dismissal periods.

UNDERSTANDING HOW YOUR EXAM IS SCORED

You can gain a better perspective about the ASE certification exams if you understand how they are scored. ASE exams are scored by an independent organization having no vested interest in ASE or in the automotive industry. With CBT exams, you will receive your exam scores immediately.

Each question carries the same weight as any other question. For example, if there are 50 questions, each is worth 2 percent of the total score. The passing grade is 70 percent. That means you must correctly answer 35 of the 50 questions to pass the exam.

Your exam results can tell you

- Where your knowledge equals or exceeds that needed for competent performance, or
- Where you might need more preparation.

Your ASE exam score report is divided into content "task" areas; it will show the number of questions in each content area and how many of your answers were correct. These numbers provide information about your performance in each area of the exam. However, because there may be a different number of questions in each content area of the exam, a high percentage of correct answers in an area with few questions may not offset a low percentage in an area with many questions.

It should be noted that one does not "fail" an ASE exam. The technician who does not pass is simply told "More Preparation Needed." Though large differences in percentages may indicate problem areas, it is important to consider how many questions were asked in each area. Since each exam evaluates all phases of the work involved in a service specialty, you should be prepared in each area. A low score in one area could keep you from passing an entire exam. If you do not pass the exam, you may take it again at any time it is scheduled to be administered.

There is no such thing as average. You cannot determine your overall exam score by adding the percentages given for each task area and dividing by the number of areas. It does not work that way because there generally is not the same number of questions in each task area. A task area with 20 questions, for example, counts more toward your total score than a task area with 10 questions.

Your exam report should give you a good picture of your results and a better understanding of your strengths and areas needing improvement for each task area.

Types of Questions on an ASE Exam

Understanding not only what content areas will be assessed during your exam, but how you can expect exam questions to be presented will enable you to gain the confidence you need to successfully pass an ASE certification exam. The following examples will help you recognize the types of question styles used in ASE exams and assist you in avoiding common errors when answering them.

Most initial certification tests are made up of between 40 and 80 multiple-choice questions. The five-year recertification exams will cover the same content as the initial exam; however, the actual number of questions for each content area will be reduced by approximately one-half. Refer to Section 4 of this book for specific details regarding the number of questions to expect to receive during the initial Heating, Ventilation & Air Conditioning (T7) certification exam.

Multiple-choice questions are an efficient way to test knowledge. To correctly answer them, you must consider each answer choice as a possibility, and then choose the answer choice that *best* addresses the question. To do this, read each word of the question carefully. Do not assume you know what the question is asking until you have finished reading the entire question.

About 10 percent of the questions on an actual ASE exam will reference an illustration. These drawings contain the information needed to correctly answer the question. The illustration should be studied carefully before attempting to answer the question. When the illustration is showing a system in detail, look over the system and try to figure out how the system works before you look at the question and the possible answers. This approach will ensure you do not answer the question based upon false assumptions or partial data, but instead have reviewed the entire scenario being presented.

MULTIPLE-CHOICE/DIRECT QUESTIONS

The most common type of question used on an ASE exam is the direct multiple-choice style question. This type of question contains an introductory statement, called a stem, followed by four options: three incorrect answers, called distracters, and one correct answer, the key.

When the questions are written, the point is to make the distracters plausible to draw an inexperienced technician to inadvertently select one of them. This type of question gives a clear indication of the technician's knowledge.

Here is an example of a direct style question:

1. When checking the coolant condition with a supplemental coolant additive (SCA) test strip, the technician finds that the coolant is severely over-conditioned. What should the technician do?

    A. Add more antifreeze to increase the SCA.
    B. Continue to run the truck until the next PMI.
    C. Drain the entire coolant system and add the proper SCA mixture.
    D. Run the truck with no SCA additives until the next PMI.

TASK C.3

Answer A is incorrect. SCA is a corrosion inhibitor additive. Adding more antifreeze does not change the SCA reading.

Answer B is incorrect. Continuing to run the truck until the next PMI will not fix the problem. It is possible that if the coolant is only slightly over-conditioned the truck could be run until the next service interval and be retested. However running a diesel engine with a severely over conditioned cooling system can cause engine damage.

Answer C is correct. The technician needs to drain the entire coolant system and add the proper SCA mixture to achieve the correct percentage of SCA.

Answer D is incorrect. Action should be taken to correct the problem as soon as possible.

COMPLETION QUESTIONS

A completion question is similar to the direct question except the statement may be completed by any one of the four options to form a complete sentence. Here's an example of a completion question:

TASKS A.2, A.4

2. During an HVAC performance test, the technician hears the A/C compressor clutch slip briefly upon engagement. The most likely cause is:

A. A worn out compressor clutch coil.

B. A defective A/C compressor clutch relay.

C. The compressor clutch air gap is too large.

D. The compressor clutch bearing is worn.

Answer A is incorrect. If the compressor clutch coil were defective, then the clutch would not engage at all.

Answer B is incorrect. A defective relay will prevent the clutch from engaging.

Answer C is correct. If the air gap is too large, the clutch may slip briefly upon engagement.

Answer D is incorrect. A worn clutch bearing may cause a loud clutch engagement, but not a slipping clutch.

TECHNICIAN A, TECHNICIAN B QUESTIONS

This type of question is usually associated with an ASE exam. It is, in fact, two true-false statements grouped together, such as: "Technician A says…" and "Technician B says…", followed by "Who is correct?"

In this type of question, you must determine whether either, both, or neither of the statements are correct. To answer this type of question correctly, you must carefully read each technician's statement and judge it on its own merit.

Sometimes this type of question begins with a statement about some analysis or repair procedure. This statement provides the setup or background information required to understand the conditions about which Technician A and Technician B are talking, followed by two statements about the cause of the concern, proper inspection, identification, or repair choices.

Analyzing this type of question is a little easier than the other types because there are only two ideas to consider, although there are still four choices for an answer.

Again, Technician A, Technician B questions are really double true-or-false statements. The best way to analyze this type of question is to consider each technician's statement separately. Ask yourself, "Is A true or false? Is B true or false?" Once you have completed an individual evaluation of each statement, you will have successfully determined the correct answer choice for the question, "Who is correct?"

An important point to remember is that an ASE Technician A, Technician B question will never have Technician A and B directly disagreeing with each other. That is why you must evaluate each statement independently.

An example of a Technician A/Technician B style question looks like this:

3. The upper radiator hose has a slight bulge. Technician A says that the hose does not need to be replaced unless it appears cracked. Technician B says that the bulge indicates a weak spot and the hose should be replaced. Who is correct?

TASK C.4

   A. A only
   B. B only
   C. Both A and B
   D. Neither A nor B

Answer A is incorrect. A bulge in the hose indicates a weak spot.

Answer B is correct. Only Technician B is correct. The hose should be replaced immediately. All of the coolant hoses should be thoroughly checked at this time. If the hoses are the same age as the failed hose, it is advisable to replace all of them at the same time.

Answer C is incorrect. Only Technician B is correct.

Answer D is incorrect. Technician B is correct.

EXCEPT QUESTIONS

Another type of question used on ASE exams contains answer choices that are all correct except for one. To help easily identify this type of question, whenever they are presented in an exam, the word "EXCEPT" will always be displayed in capital letters. Furthermore, a cautionary statement will alert you to the fact that the next question is different from the ones otherwise found in the exam. With the EXCEPT type of question, only one incorrect choice will actually be listed among the options, and that incorrect choice will be the key to the question. That is, the incorrect statement is counted as the correct answer for that question.

Be careful to read these question types slowly and thoroughly; otherwise, you may overlook what the question is actually asking and answer the question by selecting the first correct statement.

An example of this type of question would appear as follows:

4. All of the following steps should be performed after an HVAC repair has been made EXCEPT:

   A. Perform thorough visual inspection.
   B. Clear diagnostic codes.
   C. Operate the system to check performance.
   D. Recover the refrigerant.

TASK D.12

Answer A is incorrect. It is a good practice to perform a thorough visual inspection after each HVAC repair. The technician should inspect all of the components, fasteners, connections, and wires to make sure that all items are in place and ready to perform at a high level.

Answer B is incorrect. The technician should always clear all diagnostic codes after each repair, in order to assure that the truck leaves the shop with a clear computer.

Answer C is incorrect. The technician should operate the system to make sure it is performing up to specifications.

Answer D is correct. The system does not need to be recovered after an HVAC repair has been made. Recovery is necessary when the A/C system needs to have the refrigerant removed to repair or replace a refrigerant system component.

LEAST LIKELY QUESTIONS

LEAST LIKELY questions are similar to EXCEPT questions. Look for the answer choice that would be the LEAST LIKELY cause for the described situation. To help easily identify these types of questions, whenever they are presented in an exam the words "LEAST LIKELY" will always be displayed in capital letters. In addition, you will be alerted before a LEAST LIKELY question is posed. Read the entire question carefully before choosing your answer.

An example of this type of question is shown here:

TASK A.3

5. Which of the following statements is the LEAST LIKELY result of a touch test performed on a normal A/C system?

    A. The compressor discharge line is hot to the touch.
    B. The line exiting the orifice tube is cold with a frost ring around it.
    C. The suction line is cold with condensation droplets on it.
    D. The line exiting the condenser is not as hot as the line entering the condenser.

Answer A is incorrect. The compressor discharge line is normally hot to the touch. Higher ambient temperature will result in hotter discharge line temperatures.
Answer B is correct. The line exiting the orifice tube should not be cold enough to have a frost ring around it. Frost at this location would indicate a restricted orifice tube.
Answer C is incorrect. The suction line should be cold and will have condensation droplets if there is any humidity present in the outside air.
Answer D is incorrect. The condenser outlet line should be about 20°F cooler than the inlet line.

SUMMARY

The question styles outlined above are the only ones you will encounter on any ASE certification exam. ASE does not use any other types of question styles, such as fill-in-the-blank, true/false, word-matching, or essay. ASE also will not require you to draw diagrams or sketches to support any of your answer selections, although any of the above described question styles may include illustrations, charts, or schematics to clarify a question. If a formula or chart is required to answer a question, it will be provided for you.

Task List Overview

INTRODUCTION

This section of the book outlines the content areas or *task list* for this specific certification exam, along with a written overview of the content covered in the exam.

The task list describes the actual knowledge and skills necessary for a technician to successfully perform the work associated with each skill area. This task list is the fundamental guideline you should use to understand what areas you can to expect to be tested on, as well as how each individual area is weighted to include the approximate number of questions you can expect to be given for that area during the ASE certification exam. It is important to note that the number of exam questions for a particular area is to be used as a guideline only. ASE advises that the questions on the exam may not equal the number listed on the task list. The task lists are specifically designed to tell you what ASE expects you to know how to do and to help prepare you to be tested.

Similar to the role this task list will play in regard to the actual ASE exam, Delmar, Cengage Learning has developed six preparation exams, located in Section 5 of this book, using this task list as a guide. It is important to note that although both ASE and Delmar, Cengage Learning use the same task list as a guideline for creating these test questions, none of the test questions you will see in this book will be found in the actual, live ASE exams. This is true for any test preparatory material you use. Real exam questions are *only* visible during the actual ASE exams.

Task List at a Glance

The Heating, Ventilation & Air Conditioning (T7) task list focuses on four core areas, and you can expect to be asked a total of approximately 40 questions on your certification exam, broken out as outlined:

- A. HVAC Systems Diagnosis, Service, and Repair (6 questions)
- B. A/C System and Component Diagnosis, Service and Repair (20 questions)
- C. Heating and Engine Cooling Systems Diagnosis, Service (6 questions)
- D. Operating Systems and Related Controls Diagnosis and Repair (8 questions)

Based upon this information, the following graphic is a general guideline demonstrating which areas will have the most focus on the actual certification exam. This data may help you prioritize your time when preparing for the exam.

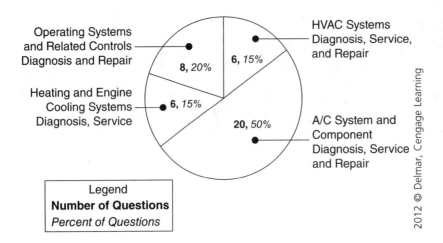

Legend
Number of Questions
Percent of Questions

2012 © Delmar, Cengage Learning

Note: There could be additional questions that are included for statistical research purposes only. Your answers to these questions will not affect your test score, but since you do not know which ones they are, you should answer all questions in the test. The five-year Recertification Test will cover the same content areas as those listed above. However, the number of questions in each content area of the Recertification Test will be reduced by one-half.

HEATING, VENTILATION & AIR CONDITIONING (T7) TASK LIST

A. HVAC Systems Diagnosis, Service, and Repair (6 questions)

1. Verify the complaint, road test the vehicle, review driver/customer interview and past maintenance documents (if available); determine further diagnosis.

A service technician's job is to try to repair the vehicle in the most efficient manner possible. To accomplish this, it is advisable to have a good strategy that is followed for each vehicle worked. The first step of repairing the vehicle is to make sure that there is a real problem. In order to verify that there is a problem, the technician needs to attempt to operate the vehicle in a manner that tests the system in question. Many times this requires that the technician road test the vehicle. After verifying that there really is something to fix, a thorough technician gathers as much information as possible to find out about the service history of the truck and also as many details about when, where, and how the problem started. Having this arsenal of information helps the technician understand the problem better and gives him/her a much higher chance of repairing the truck on the first visit.

2. Verify the need for service or repair of HVAC systems based on unusual operating noises; determine appropriate action.

The service technician must be aware of normal HVAC system operating noises in order to determine whether a system requires service. Normal noises include the sounds of A/C compressor clutch engagement, the blower motor, moving blend air and mode doors, and pressure equalization after the vehicle is shut down. Noises that could indicate the need for service include a growling sound from the water pump or A/C compressor, a whistling noise under the dash, and a grinding noise when control levers are moved.

A loose, dry, or worn A/C compressor belt will cause a squealing noise. This noise will be worse during acceleration. Worn or dry blower motor bearings may cause a squealing noise when the blower is running; this noise may occur when the engine first starts after sitting overnight. A loose or worn clutch hub or loose compressor mounting bolt will also cause a rattling noise from the compressor.

If liquid refrigerant enters the compressor, a thumping, banging noise will result. Heavy knocking compressor noises come from refrigerant system blockage, incorrect pressures, or internal damage. A worn compressor pulley bearing or clutch bearing will cause a growling noise with the compressor disengaged. If the growling noise only occurs when the system engages the clutch, internal bearings may be at fault.

3. Verify the need of service or repair of HVAC systems based on unusual visual, smell, pressures, and component temperature conditions (non-contact thermometer); determine appropriate action.

The service technician must be aware of abnormal conditions in the HVAC system in order to determine the need for system service. If the driver complains of high or low temperatures inside the cab, this is cause for a system performance test.

It is wise for a technician to pay attention to the sights, sounds, and smells of the truck when repairing trucks. The characteristics of a normal operating system include the following:

- Lines, hoses, and components on the high side of the A/C system will normally be warm or hot.
- Lines, components, and hoses on the low side of the A/C system will normally be cool or cold, with the suction line being the coldest section that can be felt or easily measured.
- The compressor clutch should make an audible noise when the A/C system is first turned on.
- Ice at any point in the A/C system is a sign of a malfunction that should be investigated.
- A musty or moldy smell that comes from the dash could be a mildew problem on the evaporator core or a stopped up HVAC duct drain tube.
- A sweet smell or a fogged up windshield is a sign of a possible heater core leak.

4. Identify system type and components (cycling clutch orifice tube—CCOT, expansion valve), and conduct performance test(s) on HVAC systems (vent outlet temperature and air flow); determine appropriate action.

Most A/C systems use a thermal expansion valve (TXV) as shown in the figure or a fixed orifice tube (FOT) in the evaporator inlet line to control refrigerant flow into the evaporator. The high-pressure liquid is on the inlet side of the expansion device and low-pressure liquid is on the outlet side. Mack trucks use a block type assembly that contains an equalized TXV. See the figure below for the various styles of TXVs available.

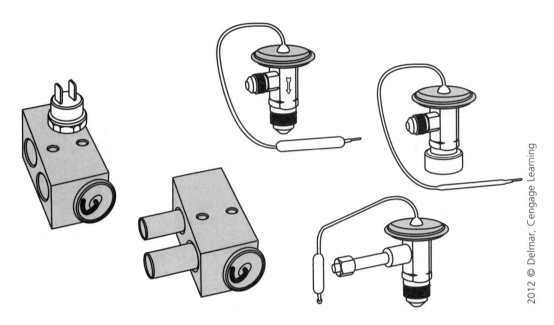

2012 © Delmar, Cengage Learning

The truck industry uses three types of A/C systems: mechanical, semi-automatic, or automatic. Mechanical systems use a slide type lever or rotary switch to control the in-cab temperature manually. In a semi-automatic system, a computer electronically controls outlet air temperature only; in automatic systems, most of the sub-systems are computer controlled.

Manual systems rely on the driver to select the temperature, mode, and blower speed. Semi-automatic temperature control (SATC) systems regulate only the temperature of the output air and rely on the driver to select the desired mode and blower speed. A microprocessor (computer) controls a fully automatic temperature control (ATC) system. These systems use the input from various sensors throughout the vehicle to control the blend doors automatically and to adjust the interior temperature using an appropriate blower speed.

An HVAC performance test should include operation of the system in all modes and at a variety of temperatures and blower speeds. A small pocket thermometer that is known to be accurate should be used to verify that the output air temperatures match the temperature settings. See Task B.1.3 for more information on conducting a performance test.

5. Identify HVAC control system type; check and record electronic diagnostic codes/indicator lights; determine further diagnosis.

HVAC control systems come in three different levels, including manual climate control, semi-automatic climate control, and fully automatic climate control. Manual climate control systems require the driver or passenger to manually select the location that the air comes out (mode), the temperature of the air, and the blower speeds. The selector knobs can be push style, pull style, or slide style. Manual climate control systems usually do not have any built in diagnostic features. The technician will need to verify that each input of the control head creates the correct output of the HVAC system. If any of the functions of the climate control head become inoperative, most control heads are serviced as an assembly.

Semi-automatic climate control systems typically have electronic temperature and blower control features, but will still have a manual mode selector. The driver or passenger can electronically choose the desired temperature for the cab; the semi-automatic climate control system will command the HVAC system to operate the devices in a way to achieve that temperature. These systems will always be integrated with one or more logic devices that are connected to the inputs required to monitor cab and outside temperatures.

Automatic climate control systems are similar to the semi-automatic systems because they possess electronic temperature and blower control features. However, these systems also control the mode of the HVAC system electronically. Both semi-automatic and fully automatic climate control systems function like any computer system by using one or more computers that are connected to several input devices. These devices supply data to the computer which is processed and results in output signals being sent to the A/C compressor, the mode doors, the temperature door, and the blower motor. See Task D.11 for more details.

B. A/C System and Component Diagnosis, Service, and Repair (16 questions)

1. A/C System—General (6 questions)

1. Diagnose the cause of temperature control problems in the A/C system; determine needed repairs.

To check for causes of temperature control problems, begin by checking for compressor clutch engagement when the A/C is selected. You should hear the clutch engage and notice a change in engine RPM and sound. If you are unable to tell by this method, watch the compressor clutch while an assistant turns the A/C on. If the clutch fails to engage, check the wiring schematic for the system to identify the power and ground sources and any compressor control devices used. Follow the manufacturer's procedures to pinpoint the source of the failure. If the compressor engages, a quick check to see if the system is operating can be made by comparing the compressor suction and discharge temperatures. The suction side should be cool to the touch and the discharge should be hot. Perform further testing by attaching a manifold gauge set and doing a performance test. Compare results to the manufacturer's specifications. After verifying that refrigerant temperatures and pressures are acceptable, check for air temperature control system problems.

The temperature of A/C system output air is generally controlled in one of two ways. In blend air systems, air cooled by the evaporator core is mixed with air warmed by the heater core. Ultimately, the output air temperature control occurs by regulating the amount of air allowed to flow through each core. In other A/C systems, temperature control is achieved by opening and closing the water control valve. When the temperature selector is in the full cold position, the water control valve totally blocks all hot coolant from entering the heater core. When the temperature selector is moved toward the hot setting, the water control valve opens to allow hot coolant into the heater core, resulting in warmer air being delivered to the cab. This type of system does not have a blend air door. In this type of system, the evaporator core and heater core are typically stacked right next to each other. A/C system outlet air temperature is affected by outside air temperature and humidity, engine coolant temperature, airflow through the condenser and evaporator, and level of refrigerant charge. Outlet air temperature may also be affected by mechanical or electrical failure of system components.

2. Identify refrigerant and lubricant type and check for contamination; determine appropriate action.

The easiest way to identify the type of refrigerant used in a given A/C system is to observe the service fittings. The Society of Automotive Engineers (SAE) standard J639 defines the size and type of service fittings for R-12 and R-134a A/C systems. R-12 service fittings have external threads. R-134a systems use quick-disconnect fittings. You cannot vent either R-12 or R-134a to the atmosphere. A refrigerant identifier machine may be connected to all types of A/C systems to determine refrigerant type and level of contamination. (Note: Recovery machines can be damaged by recovery of chemicals not compatible with the specific machine's seals and valves.)

All R-12 systems use mineral-based refrigerant oil to lubricate the compressor and prevent internal corrosion of components. Mineral-based oil is not compatible with R-134a systems. R-134a systems use polyalkylene glycol (PAG) oil. The PAG lubricant is synthetic oil and is not compatible with R-12 systems. Many retrofit kits come with ester oil included. If the technician is unable to extract all of the mineral oil during the retrofit process, then the ester oil can be used instead of the PAG oil. PAG oil and ester oils are available in several different viscosities, so the technician should only use the type recommended by the manufacturer. Installing the wrong viscosity of refrigerant oil can cause compressor failure.

3. Diagnose A/C system problems indicated by pressure gauge readings and sight glass/moisture indicator conditions (where applicable); compare gauge readings to ambient temperature/pressure chart; determine needed service or repairs.

In a normally operating A/C system, the low-side pressure varies between 20 and 45 psi, and the high-side pressure varies between 120 and 300 psi, depending on the ambient temperature and humidity levels The following table summarizes abnormal A/C system pressures and common causes.

A/C Pressure Diagnosis

LOW-SIDE PRESSURE	HIGH-SIDE PRESSURE	POSSIBLE CAUSES
LOW	LOW	Low refrigerant charge
LOW	LOW	Obstruction in the suction line
LOW	LOW	Clogged orifice tube
LOW	LOW	TXV valve stuck closed*
LOW	LOW	Restricted line from the condenser to the evaporator*
LOW	HIGH	Restricted evaporator air flow
HIGH	LOW	Internal compressor damage
HIGH	HIGH	Refrigerant overcharge
HIGH	HIGH	Restricted condenser air flow
HIGH	HIGH	High engine coolant temperature
HIGH	HIGH	TXV valve stuck open
HIGH	HIGH	Air or moisture in the refrigerant

*Stuck closed TXV valves or a restricted line from the condenser to the evaporator will cause frosting at the point of restriction.

In some A/C systems, a sight glass allows the service technician to make a quick assessment of the system condition. With the A/C compressor clutch engaged, a properly charged system will occasionally show traces of bubbles. A sight glass that appears foamy indicates that the refrigerant charge is low. When the sight glass contains bubbles and/or foam, the refrigerant charge is low and air has entered the system. Oil streaks appearing in the sight glass indicate that compressor oil is circulating through the system. A cloudy sight glass indicates that the desiccant pack in the receiver/drier has broken down.

On R-12 systems, sight glass indications are only valid when the ambient (surrounding area) temperature is above 71°F (21°C). If the temperature is below 70°F, it is normal for bubbles to appear in the sight glass. A clear sight glass may indicate the proper refrigerant charge. A clear sight glass may also indicate an excessive refrigerant charge or no refrigerant charge. Many R-134a systems do not have a sight glass. Most manufacturers agree that the following conditions must be present when performance testing an R-134a system:

- Temperature control set in the lowest position
- High blower speed
- Engine speed set at 1500 rpm or peak governed RPM
- Re-circulation air set
- Low-side and high-side pressure monitor
- Temperature of the air from the center duct monitored

Most manufacturers have eliminated the sight glass as a diagnostic means due to its unreliability when used in R-134a visual interpretations. Sometimes it is left in place in the system if the manufacturer did not wish to redesign that section, but is covered by tape. Leave the tape in place and disregard the sight glass.

4. Perform A/C system leak test; determine needed repairs.

Refrigerant systems use the following two methods to detect leaks: dye check or electronic leak detector. To check an A/C system for leaks, the technician must first ensure that the system contains enough refrigerant to allow compressor clutch engagement. If the system is empty, you should first pressure test the system using nitrogen. When you are sure the system will hold pressure, install a partial refrigerant charge. When the partial charge is installed, a small amount of dye should also be added. After running with the dye installed for 15 minutes, the dye will appear at the leak area. Ultraviolet dye is available for installation into the refrigerant and is visible using a black light detector. Electronic detectors provide an audible beeping sound when the probe is placed near the leak source. Since R-12 and R-134a are different chemically, a specific electronic detector, or one that works with both systems, is used. When checking for leaks, place the leak detector probe directly below each fitting and each component, directly below the evaporator drain, and at the center panel duct. Check the entire system to rule out multiple leaks.

When using an electronic leak detector, calibrate the detector before each use. Another common means of leak detection is a soap/water solution applied to external areas of the system. If leaks are present under pressure, then bubbles will appear. Yet another method is fluorescent soap; as with the dye, it will change color when it comes into contact with the refrigerant.

Under no circumstances is a flame-type tester to be used. When R-12 comes into contact with a flame it becomes phosgene gas. Also, all refrigerants will combust at the appropriate temperature.

5. Recover A/C system refrigerant; determine amount of oil removed; determine appropriate action.

Recovering refrigerant from an A/C system requires the use of a certified machine to pull the refrigerant from the A/C system. It is against the law to knowingly vent refrigerant to the atmosphere. Most certified recovery machines will pull the recovered refrigerant through a filter and then store it in a sealed container. In addition, many recovery machines are equipped with very accurate scales that can be used to weigh the amount of refrigerant that is removed during the recovery process. This information is valuable for the technician to find out if the A/C system was low on refrigerant. The technician should also monitor the oil container on the recovery machine before and after recovering the refrigerant in order to see how much oil was removed during the process. The technician should always add the same amount of oil that was removed when recharging the A/C system. Most recovery machines also have a recharging function that allows the machine to pull a vacuum on the A/C system and then add the correct amount of refrigerant back into the A/C system.

6. Evacuate A/C system using appropriate equipment.

Before technicians can evacuate an A/C system, they must first recover any remaining refrigerant. It is important to follow the manufacturer's instructions for the specific recovery station used. Never vent refrigerant into the atmosphere: It is illegal and environmentally irresponsible.

When the system is completely empty, connect the manifold gauge center hose to a vacuum pump. Operate this pump for 30 minutes with the service valves open and the low-side gauge valve open. After 5 minutes of operation, the low-side gauge should indicate 20 in. Hg (67.6 kPa), and the high-side should read below zero unless it is restricted by a stop pin. If the high-side gauge does not drop below zero, this indicates refrigerant blockage. When the technician locates a blockage, it must be repaired before proceeding with the evacuation process. After 15 minutes of evacuation, the low-side should indicate 24 to 26 in. Hg (81–88 kPa) if there are no leaks. If less than this value, close the low-side gauge valve and observe the gauge. If the low-side gauge needle rises slowly, this indicates a refrigerant leak. Fix the leak and proceed to evacuate the system to at least 27 in. Hg (91.4 kPa). Most manufacturers recommend that the vacuum pump run for at least 30 minutes to ensure that all moisture is removed from the system.

7. Internally clean contaminated A/C system components and hoses.

If desiccant pack deterioration or catastrophic compressor failure occurs, clean all refrigeration system components internally. Internal cleaning of A/C system components is best accomplished by flushing with A/C flush solvent. Before flushing, the compressor and all restricting components and filters must be removed from the system. It is important to regulate the pressure from the nitrogen supply tank to normal system pressure for each component.

Rather than system flushing, many truck manufacturers recommend using an in-line filter between the condenser and the evaporator to remove debris. These in-line filters come with or without a fixed orifice tube (FOT). If you use the filter that contains a FOT, you must remove the other FOT from the system.

8. Charge A/C system with correct type and quantity of refrigerant and lubricant.

The technician must complete the recovery and evacuation procedure before charging a refrigerant system. Modern recovery and charging stations do not require the A/C system to operate during system charging. It is always best to read all of the manufacturer's instructions for the specific charging station used. The original equipment manufacturer (OEM) may recommend high-side (liquid) or low-side (vapor) charging procedures. You must close both the high-side and low-side manifold gauge valves.

Connect the center hose to the proper refrigerant container and open the container valve. With the engine not running and using the high-side (liquid) charging process, open the high-side gauge valve and observe the low-side gauge, then close the high-side gauge. If the low-side gauge does not move from a vacuum to a pressure, this means the refrigerant system is restricted. With no restriction present, open the high-side gauge valve to proceed with the high-side (liquid) charging procedure. Charging is complete when the correct weight of refrigerant has entered the system. Turn the compressor over by hand to make sure that no liquid refrigerant is in the compressor. Now start the engine and run an A/C performance test. If the system is charged with the engine running and the A/C turned on, make sure to charge only through the low side.

9. Recycle refrigerant.

Differences among the terms recover, recycle, and reclaim must be completely understood and properly used within the industry. To recover refrigerant is to remove refrigerant in any condition from a system and store it in an external container. The refrigerant must then either be recycled on-site or shipped off-site for reclamation. To recycle refrigerant is to reduce contaminants in used refrigerant by separating the oil with single or multiple passes through devices, such as replaceable filter driers, which reduce moisture, acidity, and particulate matter. To reclaim refrigerant is to reprocess refrigerant to new product specifications by means which may include distillation. Chemical analysis of the refrigerant is required to assure that appropriate product specifications are met. Reclamation usually implies the use of procedures available only at specialized processing or manufacturing facilities.

10. Handle, label, and store refrigerant.

Any portable container used for transferring reclaimed or recycled refrigerant must conform to US Department of Transportation (DOT) and Underwriters Laboratories (UL) standards. Before introducing refrigerant into an approved storage cylinder, the cylinder must be evacuated to at least 27 in. Hg (91.4 kPa). The cylinder-safe filling level must be monitored by measured weight. Shut-off valves are required within 12 inches. U (30 cm) of service hose ends. Shut-off valves must remain closed while connecting and disconnecting hoses to vehicle air conditioning service ports. Safety goggles should always be worn while working with or around refrigerant.

Refrigerant should only be stored in a stable area that is not close to a heat source.

11. Test A/C system and recycled refrigerant for non-condensable gases.

To test a refrigerant for non-condensable gases, compare the pressure of the refrigerant in a cylinder to the theoretical pressure of pure refrigerant at a given temperature. If the actual pressure in the cylinder is higher than the theoretical pressure, the refrigerant is contaminated with non-condensable gas.

▦ 12. Maintain and verify correct operation of certified equipment.

The Clean Air Act (CAA) establishes the following rules for record keeping and operation of certified refrigerant service equipment:

1. Any person who owns approved refrigerant recycling equipment certified under the Act must maintain records of the name and address of any facility to which refrigerant is sent.
2. Any person who owns approved refrigerant recycling equipment must retain records demonstrating that all persons authorized to operate the equipment are certified under the Act.
3. Public Notification: Any person who conducts any retail sales of a Class I or Class II substance must prominently display a sign that reads: "It is a violation of federal law to sell containers of Class I and Class II refrigerant of less than 20 pounds of such refrigerant to anyone who is not properly trained and certified."
4. Any person who sells or distributes any Class I or Class II substance that is in a container of less than 20 pounds of such refrigerant must verify that the purchaser is certified, and must retain records for a period of three years. (These records must be maintained on-site.)

2. Compressor and Clutch (5 questions)

▦ 1. Diagnose A/C system problems that cause protection devices (pressure, thermal, and electronic) to interrupt compressor operation; determine needed repairs.

A variety of A/C system protection devices can be used in mobile A/C systems. The low-pressure cut-out switch will interrupt compressor operation if system pressure drops to the point that a loss of refrigerant charge to the compressor occurs. The high-pressure cut-out switch interrupts compressor operation in case of extremely high system pressure. The binary switch combines the function of the low- and high-pressure cut-out switches. Some systems have a high-pressure relief valve mounted in the receiver/dryer. This valve opens and relieves system pressure if the pressure exceeds 450 to 550 psi (3100 to 3792 kPa). Condenser airflow restrictions cause these extremely high pressures. In gasoline and diesel engine electronic fuel management systems, the computer operates a relay that supplies voltage to the compressor clutch. All input signals go to the engine computer. In some applications, this includes a refrigerant pressure signal. If this input signals an abnormally low- or high-pressure condition, the engine computer will not engage the compressor.

Cycling clutch orifice tube (CCOT) systems use a pressure cycling switch to cycle the compressor off and on in relation to low-side pressure. This switch is mounted in the accumulator between the evaporator and the compressor. This switch closes and turns on the compressor when the refrigerant pressure is above 46 psi (315 kPa). The dash A/C switch supplies the power to the cycling switch. The pressure switch opens when the system pressure decreases to 25 psi (175 kPa). This cycling action maintains the evaporator temperature at 33°F (1°C).

Some refrigerant systems use a thermostatic clutch cycling switch that cycles the compressor on and off in relation to evaporator outlet temperature.

2. Inspect, test, and replace A/C system pressure, thermal, and electronic protection devices.

A/C system pressure protection switches are normally closed and may be tested by checking for continuity using a digital voltmeter (DVOM) or a circuit tester while the system is at normal pressure and temperature conditions. The high-pressure switch opens if air conditioning pressure exceeds about 425–435 psi to prevent damage and closes when pressure drops to below 200 psi. The normally closed low-pressure switch, which is usually located on the accumulator, opens when the low side pressure drops below 20–25 psi. Some systems combine several functions into either a binary or trinary switch. Other systems have a thermal fuse, which blows when the compressor begins overheating. Compressor head temperature switches, when used, are mounted so that they contact the compressor case; when they sense that the temperature is exceeding a certain threshold they will open, turning the compressor clutch off. There is usually a diode, which is placed across the compressor clutch coil to reduce voltage spikes that are caused when the clutch is cycled. This is to protect other voltage-sensitive devices. If it is faulty, it will not prevent the clutch from engaging, but will fail to protect other voltage-sensitive parts of the electrical system.

3. Inspect and replace A/C compressor drive belts, pulleys, idlers and tensioners, mountings and hardware; adjust drive belts and check alignment.

When inspecting an A/C system, it is important that the technician not overlook the A/C compressor drive belts and pulleys. Drive belt edge wear indicates a misaligned or bent pulley. If the belt is loose or bottomed out in the pulley, the belt may slip and cause inadequate cooling. Cracked, glazed, or frayed belts must be replaced. Use a standard drive belt tension gauge to check and adjust belt tension. An improperly adjusted drive belt will wear or fail prematurely. A drive belt adjusted too loosely may slip and cause belt squealing, especially on acceleration with the A/C on and clutch engaged. A drive belt adjusted too tightly may cause internal engine wear or damage to other belt-driven components. Cracked or bent pulleys must be replaced, not repaired.

Damaged A/C compressor mounts or mounting plates can cause drive belt misalignment, improper drive belt tension, and compressor vibration. Welding can repair cracked mounts and mounting plates. Care must be taken to align all parts properly, however, as serpentine belts require alignment of components to within one degree.

The technician should closely investigate the compressor bracket and mounting bolts each time that an increased noise level is detected at the compressor. It is very easy to misdiagnose a compressor noise as a faulty compressor when the compressor bracket is actually the problem. Another possible source for A/C system noise is a discharge line rubbing a metallic engine component. The normal routing of the discharge line prevents this line from touching anything metallic.

4. Inspect, test, service, and replace A/C compressor clutch components or assembly.

The A/C compressor clutch assembly allows the A/C compressor to engage and disengage to modulate system pressures. The components of the compressor clutch assembly are:

- Driven plate—keyed to the compressor driveshaft
- Drive plate—integral to the drive pulley
- Clutch bearing—functions when the clutch is disengaged
- Clutch coil—creates the magnetic field that engages the clutch

Defective compressor clutch bearings will make a growling noise with the engine running and the clutch disengaged. Test the compressor clutch by applying power and ground to the appropriate terminals and watching for clutch engagement. I inspect the pulley and armature plate frictional surfaces for wear and oil contamination. Check the hub bearing for roughness, grease leakage, and looseness. A driven plate that drags on the drive plate or slips briefly on engagement indicates an improper clutch air gap. Adjust this air gap using shims: Remove shims to decrease clearance and increase shims to increase clearance. Check the gap on any clutch service. A technician can perform a voltage drop test while the system is in operation. This is the most practical test method, as components are in actual running mode.

5. Inspect and correct A/C compressor lubricant level.

Check the A/C compressor lubricant level and adjust it any time there is evidence of lubricant loss from the system. An excessive amount of oil in a refrigerant system reduces system cooling efficiency. To check the compressor lubricant level, remove the compressor from the vehicle, drain all refrigerant oil, and refill it to the manufacturer's specifications.

R-12 systems require a mineral oil with a YN-9 designation. R-134a systems with a reciprocating compressor must have a synthetic (PAG) oil designation. Rotary compressors use a different type of synthetic oil. If the oils used become mixed, compressor damage will result.

6. Inspect, test, and replace A/C compressor.

A technician can diagnose A/C compressor internal damage using a standard A/C gauge set. Low high-side pressure and high low-side pressure on the manifold gauge set may indicate a defective compressor. A faulty compressor bearing will make a growling noise with the engine running and the compressor clutch engaged. Oil dripping from the front of the compressor indicates a faulty front seal. A rattling noise may be caused by loose compressor mounts or loose compressor mounting bolts. A defective pulley bearing will also cause a growling noise with the clutch disengaged. When replacing A/C seals or o-rings, pre-lubricate them with mineral-based refrigerant oil.

Damaged A/C compressor mounts or mounting plates can cause drive belt misalignment, improper drive belt tension, and compressor vibration. Welding can repair cracked mounts and mounting plates; however, care must be taken to align all parts properly, as serpentine belts require alignment of components to within one degree.

The technician should closely investigate the compressor bracket and mounting bolts each time that an increased noise level is detected at the compressor. It is very easy to misdiagnose a compressor noise as a faulty compressor when the compressor bracket is actually the problem. Another possible source for A/C system noise is a discharge line rubbing a metallic engine component. The normal routing of the discharge line prevents this line from touching anything metallic.

3. Evaporator, Condenser, and Related Components (5 questions)

1. Correct system lubricant level when replacing the evaporator, condenser, receiver/drier or accumulator/drier, and hoses.

It is important to remember to add the correct amount of the correct viscosity of refrigerant oil when replacing the components of the A/C system. The technician needs to be cognizant of the total oil charge amount for the complete system. It is vital that refrigerant oil be installed at various points throughout the refrigeration system in order to promote a continuous flow of oil to the A/C compressor. Typically, approximately one to two ounces of oil are added to the evaporator, condenser, and receiver/drier or accumulator/drier. The technician should follow the manufacturer's recommendations when installing oil to replacement components.

2. Inspect, repair, or replace A/C system hoses, lines, filters, fittings, service ports, o-rings, and seals.

The technician should check A/C system hoses for damage and leaks during the course of any A/C system maintenance or inspection. Hoses should be replaced if they are cracked, kinked, abraded, or if the fittings show any signs of abuse. Disassemble leaking fittings and replace the o-rings. New o-rings should be lubricated with the appropriate refrigerant oil. Some manufacturers recommend that mineral oil be used to lubricate the o-rings. Using PAG oil on o-rings will lead to early failure due to the PAG attracting moisture to the joint. It is important to use the high-grade neoprene o-rings on R-134a systems.

A/C system filters and screens are used to prevent particulate matter (from corrosion, compressor failure, or desiccant breakdown) from circulating through the A/C system and must be replaced if they are clogged, restricted, or damaged. Some systems have a filter in the line between the condenser and the evaporator and some of these filters contain an orifice tube. You must install this type of filter in the proper direction. Special tools are required to service the spring lock couplings that are used on some hose connections.

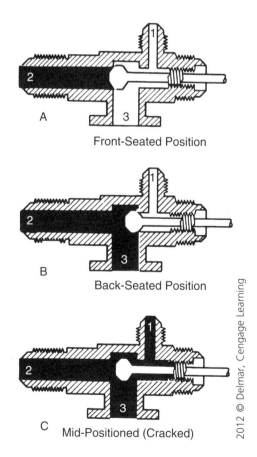

A Front-Seated Position

B Back-Seated Position

C Mid-Positioned (Cracked)

2012 © Delmar, Cengage Learning

Two types of A/C system service valves are used in mobile A/C systems. Older systems use a three-position stem-type valve. In the figure above, B shows the normal operation, or back-seated position, in which the valve stem is rotated counterclockwise to seat the rear valve face and seal off the service gauge port. C shows the mid-position used during A/C system diagnosis and service. A shows the front-seated position (valve stem rotated clockwise to seat the front valve face) that isolates the compressor from the A/C system. This position allows a technician to service the compressor without discharging the entire system. The system must never be operated with either service valve in the front-seated position.

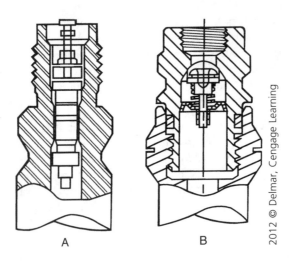

A   B

2012 © Delmar, Cengage Learning

Most R-12 A/C systems have service ports as in A in the figure above. These valve feature external threads and replaceable Schrader valves like the ones used in tire valve stems. R-134a service ports on the other hand, have internal threads and are a quick-disconnect design like figure B. The low-side port is 13 mm in diameter and the high-side port is 16 mm in diameter.

The service ports on the air conditioning system sometimes require service. These ports typically have check valves that can cause leaks at times. The repair process involves either removing the threaded valve core or removing the whole service fitting, depending on the design. The refrigerant would need to be recovered before replacing either style of check valve. After replacing the check valve, the air conditioning system would need to be evacuated and recharged with refrigerant.

3. Inspect A/C condenser for proper air flow.

The A/C condenser should be checked for proper airflow at regular intervals. During a normal A/C system inspection, any bent condenser fins should be straightened and any debris should be cleaned from the condenser. Debris in the condenser air passages causes excessive high-side and low-side pressures and reduced cooling. This problem may also cause the high-pressure relief valve to discharge refrigerant. Additionally, the technician should check the radiator shutter system for proper operation.

4. Inspect, test, and replace A/C system condenser and mountings.

If any refrigerant tubes are kinked, cracked, or leaking, the A/C condenser must be replaced. Frost on any of the condenser tubing indicates a refrigerant passage restriction. This condition results in excessive high-side and low-side pressures and inadequate cooling. If the condenser is replaced, drain the new component and install fresh refrigerant oil to manufacturer's specifications (typically one ounce). Condenser mounts and insulators should be checked for proper alignment and deformation, which could cause abrasion and fatigue damage.

5. Inspect and replace receiver/drier or accumulator/drier.

Most mobile A/C systems use a receiver/drier or an accumulator/drier to ensure an adequate supply of clean, dry refrigerant to the system. If the receiver/drier inlet and outlet pipes have a significant temperature difference, the receiver/drier is restricted. Frost forming on the receiver/drier indicates an internal restriction. Bubbles and foam in the sight glass indicate rust and moisture contamination. Both of these devices contain a bag of desiccant designed to absorb and hold traces of moisture from the refrigerant. The accumulator is located at the outlet of the evaporator. The receiver/drier is located just upstream of the system expansion device. The accumulator/drier or receiver/drier must be replaced if the A/C system has remained open to the atmosphere for an extended period of time. Other reasons for replacing the drying device are evidence of moisture or corrosion in the system, or if catastrophic compressor failure has occurred. When the accumulator/drier or receiver/drier is replaced, fresh refrigerant oil must be added to manufacturer's specifications (typically one ounce).

6. Inspect, test, and replace cab or sleeper refrigerant solenoid, expansion valve(s), thermostatic switch (thermistor); and check placement of thermal bulb (capillary tube).

One type of A/C system expansion device is the thermal expansion valve. The thermal expansion valve senses evaporator temperature using a capillary tube connected to a thermal bulb. As the fluid inside the thermal bulb expands, the orifice in the expansion valve opens to increase refrigerant flow through the evaporator. If the evaporator core temperature drops to near the freezing point, the fluid in the thermal bulb contracts and the expansion valve closes to restrict refrigerant flow. Different designs place this thermal bulb either embedded in the evaporator fins or affixed to the evaporator outlet with insulating tape.

In some systems, the expansion valve is housed in a combination valve or an "H" valve. In these systems, internal mechanisms monitor evaporator inlet and outlet temperatures and pressures and adjust the valve opening accordingly. A technician can diagnose a faulty expansion valve by using a set of A/C pressure gauges and consulting a diagnostic chart.

Some A/C systems use a thermostatic switch to monitor evaporator temperature. If the temperature gets too cold, this switch opens to interrupt power to the A/C compressor clutch. These switches are used as de-icing devices in the A/C system.

7. Inspect and replace orifice tube.

Cycling clutch-type A/C systems often use a fixed orifice tube as an expansion device. The orifice tube is located at the evaporator inlet and contains a fine screen to prevent the circulation of particulate matter through the evaporator core and back to the compressor. The orifice tube should be replaced if the screen is restricted or corroded, in case of desiccant bag breakdown, or if there is catastrophic compressor failure.

A restricted orifice tube may cause lower than specified low-side pressure, frosting of the orifice tube, and inadequate cooling from the evaporator. If these conditions appear, place a shop towel soaked in hot water around the orifice tube. If the low-side pressure increases, there is moisture freezing in the orifice tube. To rectify this condition, the technician must recover, evacuate, and recharge the system. If the hot shop towel did not increase the low-side pressure, clean or replace the orifice tube.

8. Inspect, test, and replace cab or sleeper evaporator core.

The A/C evaporator core is located (along with the heater core and airflow control doors) in the evaporator case. A technician can detect a leaking evaporator core most easily by taking a measurement at the evaporator case drain, but it may also be detected at the panel and defroster vents. Also, you can remove the blower resistor assembly and take a measurement through that cavity. If the evaporator core has a large leak, an oily film appears on the inside of the windshield and the cab temperature becomes warmer than specified. If an evaporator core is replaced, then add fresh refrigerant oil to manufacturer's specifications (typically three ounces).

9. Inspect, clean, and repair evaporator housing and water drain; check for proper evaporator air flow; inspect and service/replace evaporator air filter.

The evaporator case or housing contains the evaporator core, the heater core, the evaporator core drain, and the blend air and mode control doors. A clogged evaporator core drain will cause windshield fogging or a noticeable mist from the panel vents. A clogged drain can normally be opened up with a slender piece of wire or with low pressure shop air. The evaporator drain should be checked during routine maintenance inspections. A cracked evaporator case can cause a whistling noise during high blower operation. Minor cracks can be repaired using epoxy-type adhesives. A mildew smell that is noticeable during A/C system operation can be rectified by removing the evaporator case and washing it with a vinegar and water solution or a commercially available cleaner. Some manufacturers make an aerosol fungicide that can be sprayed onto the evaporator core surface. Access to the evaporator core can sometimes be gained by removing the blower resistor.

Many late-model trucks use a cabin air filter to clean the air inside the cab area This filter is located in HVAC duct assembly near the blower motor. This filter should be serviced periodically, depending on the environment in which the truck operates. A truck that operates under dusty conditions will require more frequent service than a truck that runs on the open road most of the time. If this filter becomes restricted, the airflow in the HVAC system will be reduced.

10. Diagnose system failures resulting in refrigerant loss from the A/C system high pressure relief device.

All mobile A/C systems are equipped with a high pressure relief device. In most systems, this device is a self-resetting relief valve which is threaded into the high side of the compressor. The high pressure relief valve will vent refrigerant from the system in the event that high-side pressure exceeds safe levels. This is often an overlooked area when total refrigerant loss has occurred.

Problems in the A/C system that could cause the pressure relief valve to vent refrigerant include:

- Refrigerant overcharge, causing excessive A/C system pressures
- Blocked airflow at the condenser, causing excessive A/C system pressures

- Inoperative cooling/condenser fan on an excessively hot day
- High-side pressure switch or sensor that fails to de-activate the compressor under very high pressures

C. Heating and Engine Cooling Systems Diagnosis, Service, and Repair (6 questions)

1. Diagnose the cause of outlet air temperature control problems in the HVAC system; determine needed repairs.

Heating system design achieves heater temperature control by one of two methods: blend air modulation or coolant flow control. In a blend air system, coolant flows through the heater core at a constant rate. Air flowing through the HVAC duct is channeled by the blend air door to flow through or around the heater core, depending on the temperature lever setting. In a coolant flow control system, temperature control is achieved by using a coolant control valve (hot water valve) to limit the amount of coolant allowed to flow through the heater core, depending on the temperature lever setting. The coolant control valve may operate by vacuum or cable.

2. Diagnose window fogging problems; determine needed repairs.

Windshield fogging may be caused by a leaking heater core or by a clogged evaporator case drain. If windshield fogging is accompanied by the smell of antifreeze, the heater core is at fault. If you smell a pungent odor, then the evaporator case drain may be clogged. A technician can usually use a stiff piece of wire to re-open a blocked evaporator case drain.

3. Perform engine cooling system tests for leaks, protection level, contamination, coolant level, temperature, coolant type, and conditioner concentration; determine needed repairs.

To test the engine cooling system for leaks, remove the radiator pressure cap following the manufacturer's instructions. Replace the cap with a standard cooling system pressure tester and pressurize the system to operating pressure. Perform a careful visual inspection for leaks. Often you can confirm cooling system contamination with a visual inspection. Rust in the system will turn the coolant an opaque reddish-brown color. If engine oil or transmission fluid has entered the system, the coolant will contain thick deposits resembling a milk shake. Coolant freeze protection level is most easily determined using a cooling system hydrometer or refractometer, which is preferred by some engine manufacturers.

To check the concentration of the cooling system conditioner, use test strips that match the type of conditioner being used and then add conditioner as needed. Be careful not to over-condition the coolant; this can lead to conditioner dropout and system damage. Cooling system temperatures are easily checked using a gauge known to be accurate or a temperature probe.

4. Inspect and replace engine cooling and heating system hoses, lines, fittings, and clamps.

Cooling system and heater hoses must be replaced if they are cracked or brittle, or if they show signs of bulging or abrasion. Hoses begin to deteriorate from the inside first, so examine the inside of any suspect hoses to determine their true condition. Often a cooling system leaks only when cold, which is known as a cold water leak. The use of constant torque clamps will minimize this. Hose clamps should be replaced if they are deformed, cracked, or cannot be operated smoothly.

5. Inspect, test, and/or replace radiator, pressure cap, and coolant recovery system (surge tank).

The technician should examine the radiator at every scheduled preventive maintenance inspection. Check the radiator for bent fins, kinked or cracked tubes, and leaks. With the engine at normal operating temperature, the temperature of the radiator core should be uniform. Cool spots indicate clogged tubes in the core. Test the pressure cap for pressure using a cooling system pressure tester. Make sure the vacuum valve in the cap is working properly. If the pressure cap seal, sealing gasket, or seat is damaged, the engine will overheat and coolant will be lost in the recovery system. Replace the cap if it does not hold the pressure specified by the manufacturer or if there is any sign of physical damage. The coolant surge tank should be checked for leaks and for sediment buildup, and should be repaired or cleaned as necessary.

If the radiator needs to be replaced, the technician should make sure that the replacement unit has sufficient cooling capacity to perform well under various conditions. The technician should replace the radiator cap, the coolant, and any hoses that appear to be weak or deteriorated. After refilling the radiator, a close inspection for leaks should be performed. The old coolant should be stored in an EPA approved container or recycled.

6. Inspect and/or replace water pump and drive system.

The water pump is driven by means of a belt drive on light-duty engines and gear driven on heavy-duty applications. The most apparent sign of water pump failure is fluid loss from the weep hole. A visual inspection of bent, misaligned, or worn pulleys must be performed and corrections made. Gear drives should be checked for clearance (lash) when reassembling. Some truck fleets replace the water pump on regular intervals to reduce the rate of failure when on long hauling trips.

7. Inspect, test, and/or replace thermostats, by-passes, housings, and seals.

One way to test a thermostat is to remove it from the vehicle, place it in a container of water with a thermometer, and heat the container until the thermostat opens. The water temperature at the point when the thermostat opens should equal the manufacturer's specification for thermostat opening temperature.

There are several signs that a technician can look for to test the thermostat. A stuck open thermostat will allow coolant to flow into the radiator when the engine is cold. In addition, a stuck open thermostat will inhibit the engine from warming up to normal operating temperature. A sign that the thermostat is stuck closed is an engine that is running hotter than normal, with an upper radiator hose that is not flowing coolant. Some truck fleets replace the thermostat on a regular basis in order to reduce the rate of failure when on long hauling trips.

It is common to replace the thermostat gasket or seal when replacing the thermostat. The engine surface as well as the thermostat housing should be carefully cleaned off to assure that the new gasket or seal will not leak.

Many technicians overlook the by-pass hose when inspecting the cooling system. This hose is more difficult to see because it is located in the front timing cover area. Always replace the by-pass hose when the other hoses are being serviced.

8. Flush and refill cooling system; bleed air from system.

The cooling system should be flushed if there is any sign of rust or contamination in the coolant. After flushing, the entire system should be drained and fresh coolant should be added. The technician should consult the service manual to verify cooling system capacity and bleeding procedure. It is a good practice to clean excess coolant from the engine and radiator after filling it up to assist in finding potential coolant leaks. In addition, it is wise to let the truck thermal cycle in order to assure that the cooling system is full. Thermal cycling involves letting the truck warm up and then letting the truck cool down again. This process helps rid the system of air pockets.

9. Inspect, test, and repair or replace engine cooling fan, hub, clutch, controls, thermostat, and shroud.

The technician should inspect the engine cooling fan for loose, cracked, or otherwise damaged blades. Also, inspect the fan hub for cracks. The cooling fan should be replaced, not repaired, if any damage is found. The cooling fan clutch may be a viscous type clutch or may be operated by a thermostatic spring. Some heavy-duty trucks use a computer-controlled clutch operated by either the chassis air system or engine oil hydraulic controls.

Since modern trucks feature fans that are controlled by a computer, an examination of fan controls must include verifying the operation of the computer input sensors such as the coolant temperature sensor and the oil temperature sensor. Also, examine any computer-controlled relays that are needed for fan operation. Use the OEM recommended electronic service tool (EST) to verify the operation of circuits controlled by the engine control module (ECM). Many trucks have a toggle switch on the dash, which allows the driver to turn the fan on manually; some systems incorporate the fan into their engine braking strategy for increased performance. Other systems use fan thermostats that have a temperature-sensing bulb immersed in the coolant, which either open or close depending upon temperature and system design. They may use either air pressure or oil pressure to actuate the fan clutch.

The fan shroud should be inspected for cracks and replaced as necessary. If fan blade and shroud damage are found, the technician should verify that the engine mounts are in good condition before replacing the shroud and blade.

10. Inspect, test, and replace heating system coolant control valve(s) and manual shut-off valves.

The coolant control valve (hot water valve) controls the flow of coolant through the heater core. The coolant control valve may be operated by vacuum or by a cable. One can verify proper valve operation by manually opening and closing the valve and observing the temperature change in the downstream heater hose. Vacuum-operated coolant valves can be tested with a hand-operated vacuum pump. A good valve should hold a vacuum for approximately one minute without leaking off. The technician should also inspect the coolant control valves for external leaks and replace them if leaks are found.

11. Inspect, check for proper air flow; flush and replace heater core.

In a poorly maintained cooling system, sediment may build up in the heater core, causing poor heater performance. If a restricted heater core is suspected, a technician can perform a temperature drop test on the unit. This test is performed by fully warming up the engine and turning the heater on full heat with the blower motor on high. The temperature is measured on the inlet and outlet heater hoses. There should not be more than an approximately 10°F drop from the inlet to the outlet.

Flushing the heater core will restore heater efficiency and may reveal small heater core leaks. The heater core can be pressure tested independently from the rest of the cooling system; however, the core should never be pressurized in excess of normal operating pressure. When replacing a heater core, it is important to reinstall all foam mounting insulators to minimize the risk of vibration damage and to ensure a good seal around the heater core.

D. Operating Systems and Related Controls Diagnosis and Repair (8 questions)

1. Diagnose the cause of failures in HVAC electrical, air, and mechanical control systems; determine needed repairs.

Diagnosing HVAC electrical control system problems is no different from diagnosing other electrical systems. The technician should follow a systematic approach to troubleshooting, including verifying the concern and performing a thorough visual inspection. The technician should use all available resources, including electrical schematics, technical service bulletins (TSBs) and electronic databases. There is no substitute for using a logical strategy when diagnosing electrical problems in the HVAC system. A technician can use a good understanding of electrical diagnosis from other truck systems and apply them to the HVAC system.

Temperature control valves and doors that direct airflow may be controlled by electric motors, vacuum motors, or air cylinders. When air cylinders are used, they are exposed to moisture and other contaminants from the truck's air system. Often they may be restored to satisfactory service by cleaning and oiling the cylinder. One truck OEM uses an air controlled water valve in the sleeper compartment that sometimes fails, resulting in air pressure leaking into the truck's cooling system.

The most common cause of failures in HVAC air and vacuum systems is leaking hoses and diaphragms. Air and vacuum leaks can often be located by listening for a hissing noise. To locate minor air leaks, brush a mild soap solution over fittings and connections and watch for bubbles. To check for vacuum leaks, use a hand-held vacuum pump to supply 20 in. Hg (67.6 kPa) to one end of the vacuum hose, while the other end is plugged or attached to its device. The hose should hold 15 to 20 in. Hg (50.8–67.6 kPa) vacuum without leaking.

2. Inspect, test, repair, and replace A/C heater blower motors, resistors, switches, relays, modules, wiring, and protection devices.

A 30 amp fuse or circuit breaker generally protects the HVAC blower circuit. Many blower systems are powered by one or more relays. The operator selects the desired blower motor speed by using the blower switch. The blower resistor block contains several resistors in series and is used to step-down the voltage to the blower motor, thereby providing multiple blower speeds. The resistor block usually contains a thermal fuse to prevent blower motor damage in case of a high current draw.

The blower resistor can be replaced by accessing it on the HVAC duct assembly. Remove the connector and then the attaching screws. Carefully install the new resistor and reconnect the screws and connector. The blower motor can be replaced by following the same method as the blower resistor. The blower motor is typically located very close to the blower resistor on the HVAC duct assembly. The blower switch is located on the HVAC control head and can be serviced individually on some units. Some manufacturers do not service the blower switch alone. The whole control head must be replaced on these units.

3. Inspect, test, repair, and replace A/C compressor clutch relays, modules, wiring, sensors, switches, diodes, and protection devices.

The A/C compressor clutch coil is usually energized by an electronically controlled relay. The compressor clutch relay may be controlled by the chassis control unit or by the engine control unit. Below is a list of electrical components that a technician may encounter when diagnosing A/C electrical problems.

- Low pressure switch: A Pressure cycling switch that opens under low A/C system pressure. It opens at 15–25 psi and closes at 35–40 psi. (two wires)
- High Pressure Switch: A pressure cut-off switch that opens under high A/C system pressure. It opens at 450–490 psi and closes at 250–275 psi. (two wires)
- Dual Pressure Switch: A binary pressure switch that opens at high and at low A/C system pressure. It is located in the high side. (three wires)
- Pressure Sensor: An A/C pressure transducer that senses pressure in the high side according to variation in the voltage level. It is located in the high side. (three wires)
- Evaporator Temperature Switch: A thermostatic switch that opens when cold temperatures are experienced in the evaporator. (two wires)
- Evaporator Temperature Sensor: A thermistor, also known as the fin temperature sensor, that senses temperature in the evaporator. It produces a variable signal. (two or three wires)
- Thermal Limiter: A device located on the compressor that interrupts power to the compressor when the temperature of the compressor gets too high.

Many manufacturers use the electrical devices outlined. Testing these devices is typically fairly simple. All pressure and thermal switches operate by opening and closing electrical contacts as the variables change. One way to test these items is to use an ohmmeter to watch the contacts open and close. When the contact is open, the ohmmeter should read out of limits (OL); when the switch is closed, the ohmmeter should read very close to zero ohms. Another way to test switches is to by-pass them with a fused jumper wire. To do this, simply disconnect the switch and then connect a fused jumper wire to the connector. If the circuit begins operating normally, then the switch that is being by-passed is open. One word of caution: Before by-passing a low pressure switch, make sure the refrigerant system is not totally empty. This can be checked by connecting a manifold set to test the static pressure.

The testing procedure is different for pressure or temperature sensors. These devices operate by varying a voltage signal as pressure or temperature changes. A digital voltmeter (DVOM) can be used to monitor signal voltage on these devices. In addition, a scan tool is a very useful diagnostic tool when testing sensors. The scan tool will display the live data from each sensor, as well as any trouble codes that might be present.

Replacing the sensors and switches is also typically an easy task. Just disconnect the electrical connector from the device and then remove the device from the vehicle. Keep in mind that pressure switches that are retained with a snap ring usually do not have a Schrader valve beneath them, so the A/C system will have to be recovered prior to removing the device.

4. Inspect and test A/C-related electronic engine and body control systems (ECM and BCM); determine needed repairs.

A/C compressor clutch operation may be dependent upon signals from various engine sensors. Defective engine-related components that affect A/C system operation can usually be diagnosed using a hand-held scan tool or a laptop PC interface. The engine control module (ECM) or body control module (BCM) will disable the A/C compressor clutch if the engine coolant temperature is too high or if the outside air temperature is too low. On a vehicle with an electronically controlled transmission, the compressor clutch can be disengaged briefly during shifts. The radiator shutter system is disabled (i.e., shutters are fully open) during A/C system operation.

5. Inspect, test, repair, and replace engine cooling/ condenser fan motors, relays, modules, switches, sensors, wiring, and protection devices.

Some trucks are equipped with electric engine cooling fans and electric condenser fans. Electric fans are usually controlled by an electronic relay, which is in turn controlled by the engine control module (ECM) or chassis management module. The ECM grounds the low- and high-speed fan relays in response to engine coolant temperature and compressor head temperature.

When the engine coolant temperature reaches a predetermined temperature set by the manufacturer, the ECM grounds the low-speed fan relay. If the coolant temperature continues to rise and exceeds another predetermined set point, the high-speed relay is grounded. The operating ranges are application specific, so check the appropriate service manual and make sure that replacement parts have the correct setting. In general, operating ranges are between 198°F and 234°F (92.2°C–112.2°C). Compressor head temperature switches, when used, are mounted so that they contact the compressor case. When they sense that the temperature is exceeding a certain threshold, they open, turning the compressor clutch off.

6. Inspect, test, repair, and replace electric and air actuator motors, relays/modules, switches, sensors, wiring, and protection devices.

Some HVAC systems control the blend air door and the mode doors with electronic actuators. Electronic actuators contain a small motor, a gear train, and a feedback device to indicate the position of the controlled door to the controlling processor. Some electronic systems contain self-diagnostic abilities with diagnostic trouble codes (DTC). The technician typically diagnoses these with a scan tool or laptop computer. On some systems, the actuator doors can be adjusted.

Automatic temperature control (ATC) systems rely on a variety of sensors to provide feedback to the ATC control unit. The control unit uses the sensor signals to determine how much heating or cooling is required to maintain the desired cab or sleeper temperature. The ambient temperature sensor monitors outside air temperature. The engine coolant temperature sensor monitors engine coolant temperature. The ATC temperature (or interior temperature) sensor monitors the temperature of the air in the cab or the sleeper systems, which automatically control the temperature in the sleeper, may be designed to minimize engine idle time.

Rather than have the engine idling during driver off-duty periods, they may shut down the engine after a predetermined idle period and then automatically restart the engine when needed in response to temperature changes in the sleeper. The parameters for idle shutdown and the high and low temperature set points may be adjusted using the appropriate electronic service tools (EST).

The electric and air actuators are typically located on the HVAC duct box and can be replaced without removing the whole duct assembly. These actuators are fastened to the duct assembly with two or three small screws. After installing new actuators, the system should be operated to assure correct operation. Electric actuators sometimes have to be calibrated after replacement.

Servicing the relays, modules, switches, sensors, wiring and protection devices requires the technician to be skilled at performing logical diagnosis and repairs. Quality wiring repairs should be performed at all times when repairing these types of circuits.

7. Inspect, test, repair, or replace HVAC system electrical, air and mechanical control panel assemblies.

Electrical control panel assemblies for manual and semi-automatic HVAC systems are modular in design, allowing for replacement of individual switches and illumination bulbs without replacing the entire panel. Most electronic ATC control panels only allow the technician access to replace illumination bulbs.

HVAC vacuum and mechanical control panel assemblies require very little testing and maintenance. A leaking vacuum switch will hiss in one or more positions. Broken control cable arms or anchors will result in ineffective control levers.

8. Inspect, test, adjust, repair, or replace HVAC system ducts, doors, outlets, control cables, and linkages.

Misaligned or improperly installed HVAC ducts will cause reduced levels of system output air. Blend air and mode control doors must be adjusted to allow a full range of motion when controls are operated.

To test the operation of the ducts, doors, and outlets, turn the blower to high and then move the selector through its full range. There is a mode door which changes the airflow source and a blend door that changes the temperature of the airflow. You should be able to hear doors open and close when you move the selectors. The output from the fan should change to the appropriate outlet soon after you hear the door open or close. If this does not happen, try to move the affected door with your hand. It may have something blocking it, such as small objects that may fall down the defroster outlets and become lodged in the lower ducts. The servo or actuating cylinder which is responsible for moving the door may also be stuck and need some lubrication. Replace any faulty actuators and retest the system. The appropriate service manual will show which position each door should be in according to various modes. When the temperature control is changed from hot to cold, the outlet temperature should also change.

Replace HVAC control cables if they are kinked or seized due to internal corrosion. Adjust control cables to allow a full range of motion for the control lever and for the output device.

Note: If poor A/C performance is evident during testing, a good first diagnostic step would be to ascertain the proper positioning and action of cable controls.

9. Diagnose constant/automatic temperature control system problems; determine needed repairs.

Most ATC systems provide internal diagnostic capabilities. On some ATC systems, diagnostic trouble codes may be displayed digitally on the control panel, while on other systems a hand-held scan tool or PC interface must be used to retrieve codes. The technician should always refer to diagnostic routines in the vehicle service manual when attempting to troubleshoot ATC diagnostic codes. Most importantly, the technician must rule out the possibility of mechanical failures before searching for electronic malfunctions.

10. Inspect, test, and replace constant/automatic temperature control microprocessor (climate control computer/programmer).

The ATC control unit may be integral to the control panel or it may be a stand-alone component. In either case, the control unit is generally multiplexed to the engine and/or body control units, thereby providing electronic diagnostic capabilities in case of control unit failure.

The battery pack should be disconnected prior to replacing the ATC computer. The technician should also touch a metal component of the truck in order to discharge the static electricity from his/her body prior to touching the computer. See task D.11 and D.12 for further information on troubleshooting this system.

11. Diagnose HVAC control system problems using on-board and/or data reader/computer to determine diagnostic codes and perform system tests; check and adjust system parameters.

The HVAC systems on late-model vehicles are typically monitored and controlled by one or more computers. The technician needs to have the ability to communicate with the computers on the vehicle. Scan tools can be used to allow the technician to communicate with the truck's computers. Scan tools can be self-contained or laptop-based. They are used to retrieve and clear diagnostic codes, view live data, and perform output function tests on the HVAC system. The scan tool is connected to the truck with a cord that hooks to the data link connector (DLC).

Semi-automatic climate control and automatic climate control systems often have a built-in diagnostic function. DTC data and calibration functions are often made available by depressing a sequence of buttons on the climate control head. The technician can depress the buttons on the climate control head to gain access to the trouble codes. The technician uses the code that is retrieved to troubleshoot the problem in the HVAC system. It is often necessary to use some type of database to retrieve troubleshooting information that can be used to diagnose and repair the truck.

12. Verify repairs and clear diagnostic codes (if applicable).

A final step to any HVAC system repair is to operate the system to make sure that it functions correctly. A thorough visual inspection at this time also is a wise step to follow. The technician should inspect all of the components, fasteners, connections, and wires to make sure that all items are in place and ready to perform at a high level. In addition to operating the system and performing a good visual inspection, the technician should clear all diagnostic codes that were present before the repair was made. The diagnostic system should be checked after the repair verification to be sure that no other problems exist at the time.

Sample Preparation Exams

Included in this section are a series of six individual preparation exams that you can use to help determine your overall readiness to successfully pass the Heating, Ventilation & Air Conditioning (T7) ASE certification exam. Located in Section 7 of this book, you will find blank answer sheet forms you can use to designate your answers to each of the preparation exams. Using these blank forms will allow you to attempt each of the six individual exams multiple times without risk of viewing your prior responses.

PREPARATION EXAM 1

1. Technician A says that reviewing past maintenance records will help determine if the truck has had past HVAC-related repairs. Technician B says that it is necessary to perform a road test to verify the problem on some trucks. Who is correct?

   A. A only
   B. B only
   C. Both A and B
   D. Neither A nor B

2. A faint hissing noise is heard from the area of the evaporator immediately after shutting down the engine with the compressor clutch engaged. Technician A says that the A/C system has a leak. Technician B says that the thermal expansion valve (TXV) is defective. Who is correct?

   A. A only
   B. B only
   C. Both A and B
   D. Neither A nor B

3. Technician A says that a weak belt tensioner can cause the drive belt to squeal momentarily as the A/C compressor engages. Technician B says that a weak belt tensioner can cause the drive belt to squeal when the engine is quickly accelerated. Who is correct?

   A. A only
   B. B only
   C. Both A and B
   D. Neither A nor B

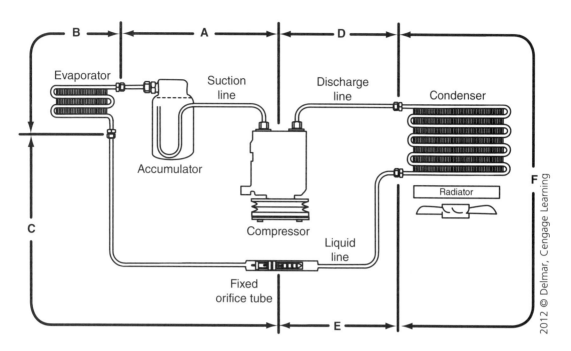

4. Referring to the figure above, which of the following conditions would most likely cause ice to form in the section labeled "F"?

    A. Air in the refrigerant system

    B. Inoperative compressor

    C. Faulty condenser fan

    D. Internal blockage in the condenser

5. The refrigerant line leading from the evaporator to the compressor contains refrigerant as a:

    A. Low-pressure gas.

    B. Low-pressure liquid.

    C. High-pressure gas.

    D. High-pressure liquid.

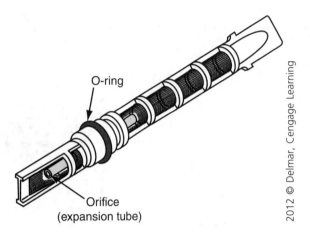

O-ring

Orifice
(expansion tube)

2012 © Delmar, Cengage Learning

6. Referring to the figure above, what function does the device shown perform for the refrigerant system?

    A. Meters refrigerant into the compressor

    B. Filters refrigerant into the condenser

    C. Meters refrigerant into the evaporator core

    D. Filters refrigerant into the compressor

7. A driver complains that he is unable to change the output temperature of his HVAC system. What is the most likely cause of this problem?

    A. Broken blend door cable

    B. Defective compressor clutch

    C. Clogged orifice tube

    D. Defective blower switch

8. A/C system pressures vary with all of the following EXCEPT:

    A. Altitude.

    B. Ambient temperature.

    C. Ambient air humidity.

    D. Cab humidity.

9. A compressor cycling on and off too fast is most commonly caused by which of the following conditions?

    A. Defective compressor clutch

    B. Defective control switch

    C. Overcharged system

    D. Low refrigerant charge

10. What will happen if R-12 comes into contact with a flame?

    A. It will explode.

    B. It will form a non-toxic gas.

    C. It will form chlorine crystals.

    D. It will form phosgene gas.

11. Technician A says that moisture in an A/C system can combine with the refrigerant to form harmful acids. Technician B says that moisture may be removed from an A/C system by evacuating. Who is correct?

    A. A only

    B. B only

    C. Both A and B

    D. Neither A nor B

12. Which of the following is the best method of removing particulate from an A/C system after a mechanical failure?

    A. Putting A/C flush solvent in the reverse flow

    B. R-11 flushing

    C. Using an in-line filter

    D. Nitrogen flushing

13. The LEAST LIKELY cause of the high-pressure relief valve tripping open is:

    A. Improper radiator shutter operation.

    B. A clogged condenser.

    C. Inoperative cooling fan clutch.

    D. A defective A/C compressor.

14. A binary pressure switch provides which of the following?

    A. Only low-pressure protection for the A/C system

    B. Only high-pressure protection for the A/C system

    C. Both low- and high-pressure protection for the A/C system

    D. Neither low-pressure nor high-pressure protection for the A/C system

15. All of the following can cause A/C compressor drive belt misalignment EXCEPT:

    A. A bent compressor mounting bracket holder.

    B. A faulty compressor clutch bearing assembly.

    C. An improperly set compressor clutch air gap.

    D. A faulty idler pulley.

16. Technician A says that the best way to ensure that an A/C compressor has the proper amount of lubricant is to drain the compressor and add lubricant to the manufacturer's specifications. Technician B says if you have a doubt about the lubricant level, add an oil charge to the system. Who is correct?

    A. A only

    B. B only

    C. Both A and B

    D. Neither A nor B

17. Which of the following items is LEAST LIKELY to be checked and inspected when replacing an A/C compressor?

    A.   The piston ring end clearance

    B.   The clutch air gap

    C.   The number of pulley grooves

    D.   The location of the mounting holes

18. The A/C condenser has several bent fins and a moderate accumulation of dead insects. Technician A says that the condenser should be replaced or it will cause the high-pressure relief valve to trip. Technician B says that the condenser fins should be straightened and the dead insects removed to optimize A/C system performance. Who is correct?

    A.   A only

    B.   B only

    C.   Both A and B

    D.   Neither A nor B

19. When attempting to verify a leaking evaporator core, the technician is LEAST LIKELY to sense refrigerant with the detector probe:

    A.   At the evaporator case drain.

    B.   Over the block type expansion valve.

    C.   At the panel vents.

    D.   At the defroster vents.

20. Which of the following examples would be the best method of removing a mildew odor from the HVAC duct?

    A.   Remove the evaporator case, clean it with a vinegar and water solution, and dry it thoroughly before reinstallation.

    B.   Spray a disinfectant into the panel outlets.

    C.   Pour a small amount of alcohol into the air intake plenum.

    D.   Place an automotive deodorizer under the dash.

21. Technician A says the R-12 A/C service valve has a removable core. Technician B says the R-134a service valve must be rear seated during A/C compressor replacement. Who is correct?

    A.   A only

    B.   B only

    C.   Both A and B

    D.   Neither A nor B

22. The A/C high-pressure relief valve shows evidence of slight oil leakage. Technician A says it is necessary to replace the valve and repair the leak. Technician B says that if the valve is just leaking a little oil and not refrigerant, repair is not necessary. Who is correct?

    A.   A only

    B.   B only

    C.   Both A and B

    D.   Neither A nor B

23. Technician A says that a clogged heater core could cause insufficient heater output. Technician B says that an improperly adjusted coolant control valve cable could cause insufficient heater output. Who is correct?

    A. A only
    B. B only
    C. Both A and B
    D. Neither A nor B

24. Which of the following tools can be used to measure the freeze protection of engine coolant?

    A. Litmus strip
    B. Manometer
    C. Refractometer
    D. Opacity meter

25. Which of the following is a typical heavy-duty truck cooling system operating pressure?

    A. 3 psi
    B. 15 psi
    C. 20 psi
    D. 30 psi

26. Each of the following is an indication that the thermostat has opened EXCEPT:

    A. Coolant flow is visible in the upper radiator tank.
    B. The upper radiator hose is hot.
    C. The lower radiator hose is hot.
    D. The temperature gauge has stabilized in the normal range.

27. Flushing the cooling system does not:

    A. Remove rust from the system.
    B. Remove contaminants from the system.
    C. Increase the life of the cooling system.
    D. Remove acids of combustion from the cooling system.

28. The inside of the windshield has a sticky film. Technician A says to check the engine coolant level. Technician B says the heater core may be leaking. Who is correct?

    A. A only
    B. B only
    C. Both A and B
    D. Neither A nor B

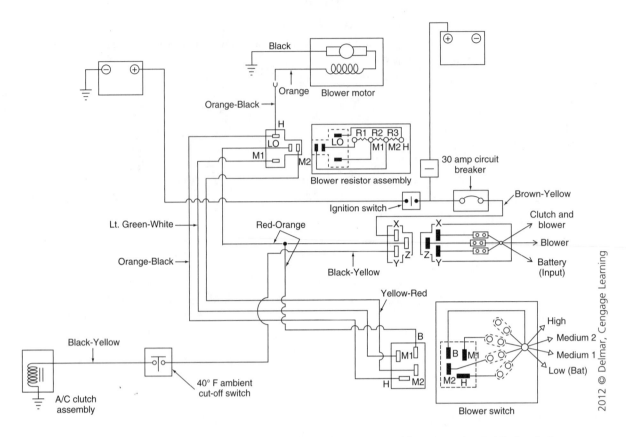

29. Referring to the figure above, all of the following could prevent the A/C clutch from engaging EXCEPT:

   A. A faulty 30 amp circuit breaker.

   B. Ambient temperature below 40° F (4.4° C).

   C. A stuck closed ambient temperature cut-off switch.

   D. An open circuit in the black-yellow wire.

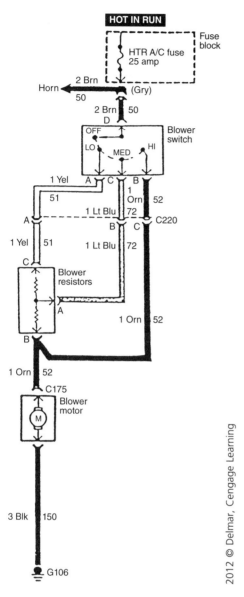

2012 © Delmar, Cengage Learning

30. Referring to the figure above, Technician A says that the light blue wire receives power when the blower switch is turned to the low position. Technician B says that the orange wire receives power when the blower switch is turned to the high position. Who is correct?

 A. A only

 B. B only

 C. Both A and B

 D. Neither A nor B

31. Technician A says that many trucks are programmed with an engine protection strategy to warn the driver if coolant level and temperature are not within preset parameters. Technician B says that imminent engine shut down can be overridden for a brief period after the alarm to allow the driver to move the vehicle to a safe place. Who is correct?

 A. A only

 B. B only

 C. Both A and B

 D. Neither A nor B

32. Technician A says that an electronic blend door motor uses a pulse-width modulated (PWM) signal to control the position of the door. Technician B says that an electronic mode door motor uses a feedback device to indicate the position of the door. Who is correct?

    A. A only
    B. B only
    C. Both A and B
    D. Neither A nor B

33. Before replacing an HVAC electronic control panel, the technician should:

    A. Remove the control cables from the vehicle.
    B. Disconnect the batteries.
    C. Disassemble the dash panel.
    D. Apply dielectric grease to the switch contacts.

34. What is the LEAST LIKELY cause of the HVAC mode switch functioning incorrectly?

    A. Vacuum leak
    B. Air leak
    C. Broken cable
    D. Open blower resistor

35. A whistling noise coming from under the dash while the HVAC system is being operated with the blower on HI could indicate which of the following?

    A. A misaligned duct
    B. A defective vacuum actuator
    C. An improperly adjusted mode door cable
    D. A poor electrical connection to the blend door motor

36. A truck is being diagnosed for an ATC system problem. The blend air door constantly moves back and forth. Which of the following conditions would be the most likely cause of this problem?

    A. Binding blend air door
    B. Defective actuator motor
    C. Improperly adjusted ATC sensor cable
    D. Defective feedback device

37. To avoid mixing refrigerants, Technician A says that the R-134a service hose fittings are different from R-12 fittings. Technician B says that R-12 containers may be either white or yellow, but R-134a containers are sky blue. Who is correct?

    A. A only
    B. B only
    C. Both A and B
    D. Neither A nor B

38. The refrigerant is being recovered from a late-model truck with an A/C recovery/recycling machine. Technician A says that some recycling machines filter the refrigerant as the recovery process is taking place. Technician B says that some recycling machines filter the refrigerant as the recharge process is taking place. Who is correct?

    A. A only
    B. B only
    C. Both A and B
    D. Neither A nor B

39. Before a portable container is used to transfer recycled R-12, it must be evacuated to at least:

    A. 20 in. Hg (67.7 kPa)
    B. 22 in. Hg (74.5 kPa)
    C. 27 in. Hg (91.4 kPa)
    D. 12 in. Hg (40.6 kPa)

40. What is the source of most non-condensable gases in refrigerant?

    A. Acid
    B. Air
    C. Moisture
    D. Oil

PREPARATION EXAM 2

1. A truck's A/C system does not cool when driving at slow speeds. Technician A says that the truck needs a fan clutch. Technician B says that the truck needs to be road tested to see if the A/C cools well at highway speeds. Who is correct?

    A. A only
    B. B only
    C. Both A and B
    D. Neither A nor B

2. Technician A says that a refrigerant identifier can be used to test for air. Technician B says that a flow tester can be used to test for air. Who is correct?

    A. A only
    B. B only
    C. Both A and B
    D. Neither A nor B

3. A hissing sound is heard under the hood area of a truck after the engine has been turned off. The noise stops after about one minute and only happens when the air conditioner or defroster has been used. Technician A says that the engine cooling system is equalizing. Technician B says that the alternator can produce this noise. Who is correct?

    A. A only
    B. B only
    C. Both A and B
    D. Neither A nor B

4. A truck technician discovers a heater core leak in the sleeper that requires replacement. Technician A says that it is important to add the proper amount of refrigeration oil before installation. Technician B says that a PAG-based lubricant is used in modern heater cores. Who is correct?

    A. A only
    B. B only
    C. Both A and B
    D. Neither A nor B

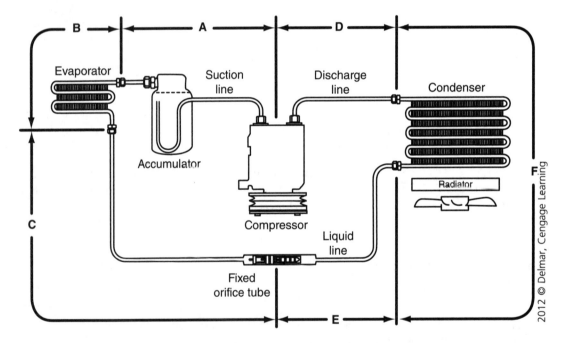

5. Referring to the figure, Section D is found to be very warm during a performance test of the A/C system on a 90° F (32.2° C) day. Technician A says that a faulty compressor clutch coil could cause this problem. Technician B says that a restricted orifice tube could cause this problem. Who is correct?

    A. A only
    B. B only
    C. Both A and B
    D. Neither A nor B

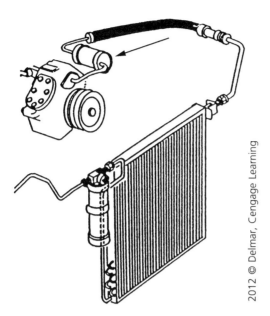

2012 © Delmar, Cengage Learning

6. Referring to the figure, if the component indicated by the arrow were not used, what would be the A/C performance result?

   A. Higher A/C high-side pressure

   B. Improperly filtered refrigerant

   C. Evaporator icing

   D. Noisier A/C operation

7. Poor cooling from an A/C system that uses a TXV valve can be caused by all of the following EXCEPT:

   A. A fan clutch that is always engaged.

   B. An improperly adjusted blend door cable.

   C. A low refrigerant charge.

   D. A refrigerant overcharge.

8. With the HVAC system in DEF mode, the blower on HI, and the temperature control on COLD, the bottom of the windshield fogs up on the outside because:

   A. The evaporator case drain is clogged.

   B. The heater core leaks.

   C. The evaporator core is iced up.

   D. The cold windshield causes moisture to condense from the outside air.

9. Where should the probe of an electronic lead detector be placed in order to most effectively find a refrigerant leak?

   A. Directly above the suspected leak area

   B. Directly below the suspected leak area

   C. Six inches upstream from the suspected leak area

   D. Six inches downstream from the suspected leak area

10. Technician A says that A/C refrigerant hoses can be flushed with R-134a. Technician B says that the condenser can be flushed with R-12. Who is correct?

   A. A only
   B. B only
   C. Both A and B
   D. Neither A nor B

11. Technician A says you can complete the high-side charging procedure with the engine running. Technician B says if liquid refrigerant enters the compressor, damage will result to the compressor. Who is correct?

   A. A only
   B. B only
   C. Both A and B
   D. Neither A nor B

12. An electric condenser fan may be controlled by any of the following EXCEPT:

   A. The engine control unit.
   B. The chassis management unit.
   C. A manual switch.
   D. An electronic relay.

13. The low-pressure cut-out switch senses pressure in the:

   A. System high side.
   B. Atmosphere.
   C. System low side.
   D. Cab/sleeper.

14. Which of the following statements about coolant control valve replacement in an ATC system is true?

   A. Refrigerant must be recovered before the coolant control valve is replaced.
   B. Heater hoses connected to the coolant control valve may be clamped during the replacement procedure to maintain coolant in the system.
   C. Access to the coolant control valve is gained through the fresh air door.
   D. The coolant control valve is an integral part of the heater hose.

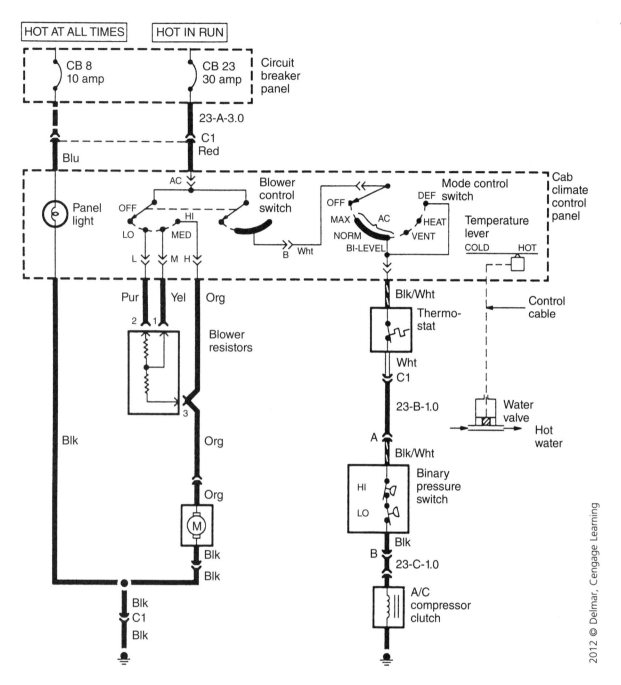

15. Referring to the figure, all of the following could prevent A/C compressor clutch engagement EXCEPT:

    A.   A defective blower switch.

    B.   An open binary pressure switch.

    C.   A poor connection at the A/C thermostat.

    D.   An open circuit breaker at CB 8.

16. The compressor discharge valve is designed to:

    A. Open after vaporized refrigerant is compressed, allowing the refrigerant to move to the condenser.

    B. Open before vaporized refrigerant is compressed, allowing the refrigerant to move to the evaporator.

    C. Regulate system variable pressure.

    D. Regulate A/C system temperature.

17. Which of the following conditions would be LEAST LIKELY to cause a knocking sound when the A/C compressor is engaged?

    A. Broken compressor mounting bolt

    B. Restricted suction line

    C. Discharge line rubbing a metal bracket

    D. Loose compressor mounting bolts

18. A signal to the engine control module (ECM) from which of the following sensors could cause the A/C compressor to disengage?

    A. Engine coolant temperature (ECT) sensor

    B. Intake air temperature (IAT) sensor

    C. Heated oxygen sensor (HO_2S)

    D. Cooling fan control sensor

19. Technician A says that some systems have an in-line filter. Technician B says that petroleum jelly should be used to lubricate new A/C o-rings. Who is correct?

    A. A only

    B. B only

    C. Both A and B

    D. Neither A nor B

20. Technician A says that the best way to test a vacuum-operated coolant control valve is to disconnect the vacuum hose and observe whether coolant flows through it. Technician B says applying and releasing vacuum and checking if the valve arm moves freely both ways is a way of testing a vacuum-operated coolant control valve. Who is correct?

    A. A only

    B. B only

    C. Both A and B

    D. Neither A nor B

21. What is the function of the orifice tube?

    A. It allows water to drain from the evaporator case.

    B. It allows refrigerant to be metered into the evaporator.

    C. It regulates refrigerant flow through the condenser.

    D. It regulates airflow through the evaporator.

22. Technician A says that some R-134a service ports have replaceable valve cores. Technician B says that a leaking R-134a service port cap will prevent refrigerant from escaping when the valve core leaks. Who is correct?

   A. A only

   B. B only

   C. Both A and B

   D. Neither A nor B

23. Which of the following is a normal range for the low-side gauge while performing a performance test?

   A. 5–10 psi

   B. 25–45 psi

   C. 60–80 psi

   D. 180–205 psi

24. When a cooling system is pressure tested, there are no obvious external leaks, but the system cannot maintain pressure. Which of the following is the most likely cause of this problem?

   A. Leaking evaporator

   B. Defective heater valve

   C. Stuck open thermostat

   D. Blown head gasket in the engine

25. Cooling and heating system hoses should be replaced for any of the following reasons EXCEPT:

   A. They leak at the hose clamps.

   B. They are cracked.

   C. They show signs of bulging.

   D. They feel spongy.

26. A growling noise is coming from the water pump. Technician A says that the water pump bearing could be the cause. Technician B says that the impeller has been eroded by cavitation. Who is correct?

   A. A only

   B. B only

   C. Both A and B

   D. Neither A nor B

27. Which of the following is the LEAST LIKELY function of the accumulator/drier that is used on orifice tube systems?

   A. Removes moisture from the refrigerant and stores it in a desiccant pouch

   B. Stores refrigerant vapor and oil

   C. Allows only refrigerant vapor and oil to be sent to the compressor

   D. Prevents vaporized refrigerant from exiting to the thermal expansion valve

28. A sweet odor comes from the panel vents while operating the A/C in NORMAL mode. Technician A says that the smell could be R-12 in the cab, so the truck must have a leaking evaporator. Technician B says the heater core could be leaking. Who is correct?

   A. A only

   B. B only

   C. Both A and B

   D. Neither A nor B

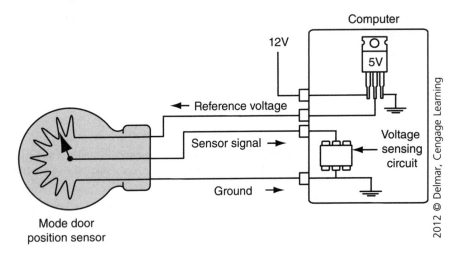

29. Referring to the figure, Technician A says that the computer sends a 5 volt reference to the mode door position switch. Technician B says that the sensor signal varies as the mode door position changes. Who is correct?

   A. A only

   B. B only

   C. Both A and B

   D. Neither A nor B

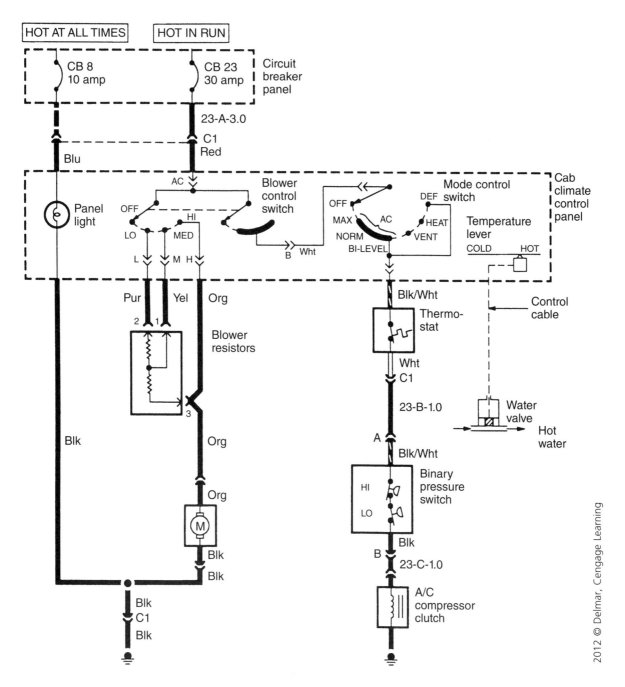

30. Referring to the figure, which of the following conditions is LEAST LIKELY to cause the blower motor to be totally inoperative?

    A. Circuit breaker 23 is open.

    B. The blower motor connector is broken.

    C. The binary pressure switch is open.

    D. The red wire at connector C1 is open.

31. Technician A says that the binary pressure switch prevents compressor operation if the refrigerant charge has been lost or ambient temperature is too cold. Technician B says that the binary pressure switch turns off the compressor if the system pressure is too high. Who is correct?

    A. A only
    B. B only
    C. Both A and B
    D. Neither A nor B

32. How much refrigerant oil should be in a typical A/C condenser?

    A. 1 ounce
    B. 5 ounces
    C. 7 ounces
    D. 11 ounces

33. Which lubricant is recommended for R-134a mobile A/C systems?

    A. Polyalkylene glycol-based (PAG-based) lubricant
    B. Mineral-based petroleum lubricant
    C. DEXRON or DEXRON II lubricant
    D. Type C-3 SAE 30 based oil lubricant

34. Which of the following is the LEAST LIKELY cause of a binding temperature control cable?

    A. A kinked cable housing
    B. Corrosion in the cable housing
    C. A deformed or over-tightened cable clamp
    D. A defective mode door

35. Which of the following conditions would most likely result from over-tightening the fasteners on an air actuator?

    A. Ruptured diaphragm
    B. Stripped threads
    C. Deformed linkage
    D. Air leak

36. A heavy truck is being diagnosed for a belt problem. What is the most likely cause for a serpentine drive belt to slip under heavy loads?

    A. Surface cracks on the inside of the belt
    B. Belt stretched 1/4 inch
    C. Seized tensioner
    D. Glazed belt

37. Technician A says that anyone who purchases R-134a must maintain records for three years indicating the name and address of the supplier. Technician B says the supplier must maintain sales records of refrigerant purchases. Who is correct?

    A. A only
    B. B only
    C. Both A and B
    D. Neither A nor B

38. Which process reduces contaminants in used refrigerant by using oil separation and filter core driers?

    A. Restoration

    B. Recovery

    C. Recycling

    D. Reclamation

39. A refillable A/C refrigerant cylinder is considered full when it reaches what capacity by weight?

    A. 50 percent

    B. 60 percent

    C. 70 percent

    D. 80 percent

40. There is a growling or rumbling noise at the A/C compressor with the compressor clutch engaged or disengaged. Technician A says the compressor bearing is defective. Technician B says the clutch bearing is defective. Who is correct?

    A. A only

    B. B only

    C. Both A and B

    D. Neither A nor B

PREPARATION EXAM 3

1. A truck is being diagnosed for an A/C problem. Technician A says that the driver's complaint should be closely inspected in order to completely understand the problem. Technician B says that a road test is never necessary when working with A/C problems. Who is correct?

    A. A only

    B. B only

    C. Both A and B

    D. Neither A nor B

2. During an HVAC performance test, the technician hears the A/C compressor clutch slip briefly upon engagement. The most likely cause is:

    A. A worn out compressor clutch coil.

    B. A defective A/C compressor clutch relay.

    C. The compressor clutch air gap is too large.

    D. The compressor clutch bearing is worn.

3. Technician A says that a heater core leak could cause a coolant smell to be present in the cab area. Technician B says that a leaking heater core could cause coolant to leak onto the floorboard area of the truck. Who is correct?

    A. A only

    B. B only

    C. Both A and B

    D. Neither A nor B

4. A cycling clutch orifice tube (CCOT) A/C system is operating at 84° F (28.9° C) ambient temperature, the compressor clutch cycles several times per minute, and the suction line is warm. The high-side gauge shows lower than normal pressures. The most likely cause of this problem could be:

   A. Low refrigerant charge.

   B. Flooded evaporator.

   C. Restricted accumulator.

   D. Overcharge of refrigerant.

5. While conducting a performance test on a semi-automatic HVAC system, a technician finds that only a small amount of air is directed to the windshield in defrost mode. The most likely cause of this problem is:

   A. A defective microprocessor.

   B. A defective blend door actuator.

   C. An open blower motor resistor.

   D. An improperly adjusted mode door cable.

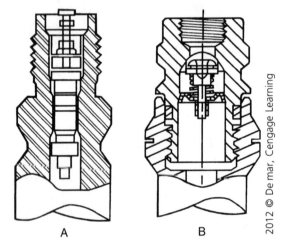

A   B

2012 © De mar, Cengage Learning

6. Referring to the figure, Technician A says that a system with service port fitting like figure A is designed to use R-134a as a refrigerant. Technician B says that the service port fitting represented by figure B is for use with R-12. Who is correct?

   A. A only

   B. B only

   C. Both A and B

   D. Neither A nor B

7. Technician A says a TXV not regulating properly could cause reduced airflow from the instrument panel outlets. Technician B says a TXV not regulating properly could cause the evaporator to ice up. Who is correct?

   A. A only

   B. B only

   C. Both A and B

   D. Neither A nor B

8. Technician A says that refrigerant with air contamination will have a higher static pressure than pure refrigerant. Technician B says that an electronic refrigerant identifier can be used to detect leak sealer additive. Who is correct?

    A. A only
    B. B only
    C. Both A and B
    D. Neither A nor B

9. A band of frost on the A/C high-pressure liquid line at a point before the orifice tube indicates:

    A. A defective compressor discharge valve.
    B. A restriction in the high-pressure hose.
    C. A clogged orifice tube.
    D. Moisture in the system.

10. When evacuating an A/C system, which manifold gauge hose is connected to the vacuum pump?

    A. The high-pressure hose
    B. The low-pressure hose
    C. The center service hose
    D. Any hose

11. All of the following statements about nitrogen flushing the A/C system are true EXCEPT:

    A. The technician should install a pressure regulator on the supply tank.
    B. The technician should disconnect the A/C compressor.
    C. The technician should remove restrictive components (i.e., STV, TXV valve) from the system.
    D. Nitrogen must not be allowed to escape into the atmosphere.

12. All of the following statements about charging an A/C system are true EXCEPT:

    A. Refrigerant may be installed through both service valves when the engine is not running.
    B. Refrigerant may be installed through the low-side service valve when the engine is running.
    C. Refrigerant may be installed through both service ports when the engine is running.
    D. You may install refrigerant directly from an approved charging station.

13. During normal A/C operation, a loud hissing noise is heard and a cloud of vapor is discharged from under the vehicle. Technician A says that the excessive high-side pressure that caused the pressure relief valve on the A/C compressor to trip may have been the result of a defective engine cooling fan clutch. Technician B says that the relief valve may have tripped due to the refrigerant system being overcharged with refrigerant. Who is correct?

    A. A only
    B. B only
    C. Both A and B
    D. Neither A nor B

14. A heavy truck with a faulty A/C pressure sensor is being diagnosed. Technician A says that this device is used to provide A/C pressure feedback to a control module. Technician B says that this device is typically mounted in the high side of the A/C system. Who is correct?

    A. A only
    B. B only
    C. Both A and B
    D. Neither A nor B

15. During a performance test, the A/C compressor clutch is observed to be slipping. Technician A says that the air gap was probably set improperly. Technician B says that the pressure plate needs to be resurfaced. Who is correct?

    A. A only
    B. B only
    C. Both A and B
    D. Neither A nor B

16. A truck has just received an HVAC repair. All of the following functions should be performed before delivering the truck back to the owner EXCEPT:

    A. Inspect the fasteners for correct torque.
    B. Recover the refrigerant.
    C. Inspect the wiring connections.
    D. Clear the diagnostic trouble codes.

17. A knocking sound is coming from the area of the A/C compressor when in operation. When the compressor is shut off, the noise stops. The A/C system cools well and there are no indications of A/C system problems. Technician A says that the noise could be caused by a broken compressor mounting bracket. Technician B says that the noise could be caused by a restricted discharge line. Who is correct?

    A. A only
    B. B only
    C. Both A and B
    D. Neither A nor B

18. Any of the following can be used to clean road debris from the condenser fins EXCEPT:

    A. A mild saline solution.
    B. A soft whiskbroom.
    C. Compressed air.
    D. A mild soap and water solution.

19. In servicing the expansion valve, which of the following can a technician perform?

    A. Adjust the expansion valve using a torque wrench.
    B. Adjust the expansion valve using an Allen wrench.
    C. Adjust the expansion valve using a screwdriver.
    D. The expansion valve cannot be adjusted.

20. Approximately how much refrigerant oil must be added to a newly replaced evaporator core?

    A. None
    B. 3 ounces
    C. 9 ounces
    D. 14.5 ounces

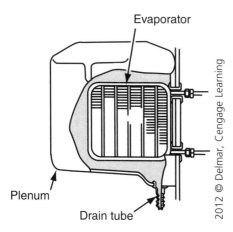

21. Referring to the figure, which of the following conditions would most likely occur if the drain tube becomes blocked?

    A. The A/C system would become colder.
    B. The heater core would become restricted.
    C. Water could leak onto the floorboard.
    D. The evaporator core would leak.

22. The A/C compressor high-pressure relief valve:

    A. Is calibrated by shimming it to the proper depth.
    B. Must be replaced if it ever vents refrigerant from the system.
    C. Will reset itself when A/C system pressure returns to a safe level.
    D. Is not used in R-134a systems.

23. Which of the following is LEAST LIKELY to cause windshield fogging in the DEFROST mode?

    A. A leaking heater core
    B. A clogged evaporator drain
    C. A water leak into the plenum chamber
    D. Moisture in the refrigerant

24. The upper radiator hose has a slight bulge. Technician A says that the hose does not need to be replaced unless it appears to be cracked. Technician B says that the bulge indicates a weak spot and the hose should be replaced. Who is correct?

    A. A only
    B. B only
    C. Both A and B
    D. Neither A nor B

25. A scan tool can be used for all of the following functions on a climate control computer EXCEPT:

    A. Calibrating the blower resistor.

    B. Retrieving trouble codes.

    C. Displaying sensor data.

    D. Displaying switch data.

26. All of these statements about cooling system service are true EXCEPT:

    A. When the cooling system pressure is increased, the boiling point increases.

    B. The boiling point decreases when more antifreeze is added to the coolant.

    C. Good quality ethylene glycol antifreeze contains antirust inhibitors.

    D. Coolant solutions should be recovered, recycled, and handled as hazardous waste.

27. When the engine cooling fan viscous clutch is disengaged:

    A. The fan blade will remain stationary.

    B. The engine idle will drop.

    C. The radiator shutters must be closed.

    D. The fan blade may freewheel at a reduced speed.

28. Technician A says that a vacuum-operated coolant control valve can be tested with a scan tool. Technician B says a vacuum-operated coolant control valve can be tested by applying and releasing vacuum and checking if the valve arm moves freely both ways. Who is correct?

    A. A only

    B. B only

    C. Both A and B

    D. Neither A nor B

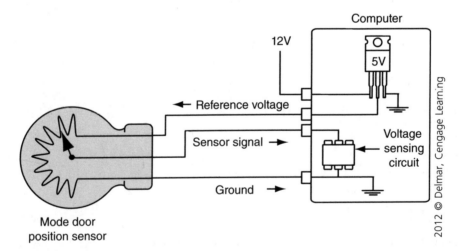

29. Referring to the figure above, which of the following readings would most likely be present on the sensor signal wire with the wiper in the current position?

    A. 0.0 volts

    B. 0.5 volts

    C. 2.5 volts

    D. 4.5 volts

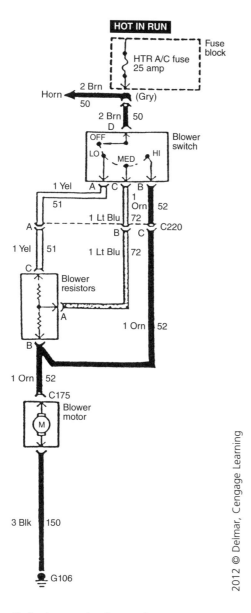

2012 © Delmar, Cengage Learning

30. Referring to the figure above, which of the following statements is LEAST LIKELY to be correct concerning the blower circuit?

A. The blower switch receives power from connector C220.

B. The blower motor would only have high speed if the blower resistor were removed.

C. The blower motor is grounded at G106.

D. The blower circuit has three speeds.

31. On accumulator-type systems with the compressor cycling switch located on the accumulator, the switch senses:

A. Outside temperature.

B. Accumulator pressure.

C. Accumulator temperature.

D. Engine compartment temperature.

32. Technician A says that the HVAC systems on most medium-duty trucks with diesel engines operate totally independently from the engine control system. Technician B says that most of the HVAC systems on these vehicles provide inputs and receive outputs from the engine control unit. Who is correct?

    A. A only
    B. B only
    C. Both A and B
    D. Neither A nor B

33. A truck with an air-controlled fan clutch is being diagnosed for a fan problem. Technician A says that some trucks have a toggle switch on the dash to enable the fan. Technician B says that the fan solenoid controls the air supply to the fan clutch. Who is correct?

    A. A only
    B. B only
    C. Both A and B
    D. Neither A nor B

34. A driver complains that the air-operated heater outputs hot air when the temperature control lever is in the COLD position. Which of the following conditions would be the LEAST LIKELY cause of this problem?

    A. A defective coolant control valve
    B. A defective engine coolant temperature sensor
    C. A defective air control solenoid
    D. A defective blend air door control cylinder

35. Inadequate airflow from one vent could be caused by which of the following conditions?

    A. Misaligned air duct
    B. Faulty blower resistor
    C. Clogged heater core
    D. High ambient humidity

36. Which of the following procedures would be the LEAST LIKELY one performed on a truck that has just been repaired by a technician?

    A. Inspect the components for leaks.
    B. Clear the diagnostic trouble codes.
    C. Identify the refrigerant.
    D. Operate the system to monitor for correct operation.

37. Owners of approved refrigerant recycling equipment must maintain records that demonstrate that:

    A. Only certified technicians operate the equipment.
    B. The equipment is operated under the supervision of certified technicians.
    C. Technicians who are undergoing certification training operate the equipment.
    D. The equipment is operated only when a certified technician is on the premises.

38. To prevent overfilling recovery cylinders, the service technician must:

    A. Monitor cylinder pressure as the cylinder is being filled.

    B. Monitor cylinder weight as the cylinder is being filled.

    C. Make sure cylinder safety relief valves are in place and operational.

    D. Occasionally shake the cylinder and observe any change of pressure while filling.

39. A truck A/C system has just been repaired. Technician A says that the system should be overcharged by 1/2 pound to account for future leaks in the refrigerant system. Technician B says that the system should be operated to make sure that it functions correctly. Who is correct?

    A. A only

    B. B only

    C. Both A and B

    D. Neither A nor B

40. Before disposing of an empty or nearly empty original container that was used to ship refrigerant from the factory, which of the following should a technician do?

    A. Clean the container and keep it for storage of recycled refrigerant.

    B. Open the valve completely and paint an X on the cylinder.

    C. Flush the container with oil and nitrogen to keep it from rusting.

    D. Recover remaining refrigerant, evacuate the cylinder, and mark it empty.

PREPARATION EXAM 4

1. Which of the following actions would be LEAST LIKELY to happen when a technician receives a repair order on a truck with an A/C complaint?

    A. Verify the complaint.

    B. Road test the truck.

    C. Review past maintenance records.

    D. Perform a heat load test on the refrigerant system.

2. A hissing sound is heard under the hood area of a truck after the engine has been turned off. The noise stops after about one minute and only happens when the air conditioner or defroster has been used. Technician A says that the A/C system is equalizing. Technician B says that the noise is not normal and the truck should be towed to the nearest repair shop. Who is correct?

    A. A only

    B. B only

    C. Both A and B

    D. Neither A nor B

3. A compressor has excessive air gap at the clutch drive plate. Technician A says that the clutch could squeal when the compressor disengages. Technician B says that the clutch could squeal when the ambient temperature is very high. Who is correct?

   A. A only
   B. B only
   C. Both A and B
   D. Neither A nor B

4. During an HVAC performance test, the Technician notices that the A/C compressor outlet is nearly as hot as the upper radiator hose. Technician A says that this is a normal condition. Technician B says that the system is overcharged. Who is correct?

   A. A only
   B. B only
   C. Both A and B
   D. Neither A nor B

5. Technician A says that the compressor clutch should not make an audible sound when engaging and disengaging. Technician B says that ice forming on the suction line is a sign of a properly functioning A/C system. Who is correct?

   A. A only
   B. B only
   C. Both A and B
   D. Neither A nor B

6. Technician A states that systems that use an orifice tube use an accumulator. Technician B states that systems that use a TXV use a receiver/drier. Who is correct?

   A. A only
   B. B only
   C. Both A and B
   D. Neither A nor B

7. In MAX A/C mode:

   A. The outside air door is closed.
   B. The defroster door is open.
   C. The compressor clutch cannot disengage.
   D. The blower is disabled.

8. An A/C system with excessive high-side pressure could be the result of all of the following EXCEPT:

   A. An overcharge of refrigerant.
   B. An overheated engine.
   C. Restricted airflow through the condenser.
   D. Ice buildup on the orifice tube screen.

9. Oil and dirt accumulation on an A/C hose connection may indicate:

    A. Excessive pressure in the system.
    B. A refrigerant leak.
    C. A defective compressor shaft seal.
    D. That there is too much oil in the system.

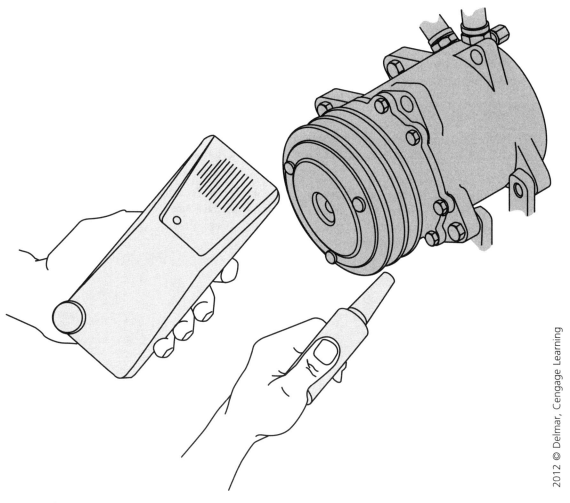

2012 © Delmar, Cengage Learning

10. What tool is the technician using in the figure above?

    A. Refrigerant identifier
    B. Belt tension gauge
    C. Refrigerant leak detector
    D. Refrigerant electronic manifold

11. Technician A says some manufacturers recommend the installation of an in-line filter between the evaporator and the compressor as an alternative to refrigerant system flushing. Technician B says an in-line filter containing a fixed orifice may be installed and the original orifice tube left in the system. Who is correct?

    A. A only
    B. B only
    C. Both A and B
    D. Neither A nor B

12. Which of these statements is correct concerning the method to recharge a heavy truck A/C system?

   A. The charging process is complete when the system reaches the correct evaporator temperature.

   B. When the low side no longer moves from a vacuum to a pressure, the process is complete.

   C. The truck engine must be running during recharging.

   D. Either a high-side or a low-side charging process can be used.

13. The A/C high-pressure switch is used to:

   A. Boost the system high-side pressure.

   B. Open the circuit to the A/C compressor clutch coil when the high-side pressure reaches its upper limit.

   C. Ensure that system pressure remains at the upper limit.

   D. Vent refrigerant from the compressor in the event of extremely high system pressure.

14. A serpentine belt is being replaced on a diesel-powered truck and the tensioner will not snap back after being released. Technician A says that the tensioner spring could be broken and the tensioner will need to be replaced. Technician B says that the idler pulley is jammed and will need to be replaced. Who is correct?

   A. A only

   B. B only

   C. Both A and B

   D. Neither A nor B

15. To check and adjust the A/C compressor lubricant level:

   A. Quickly purge the system and add oil charges to refill it.

   B. Open the drain plug and crank the engine until the compressor is empty, then pump fresh oil into the compressor.

   C. Remove the compressor from the vehicle, drain the oil, and add the specified quantity of fresh oil.

   D. Add refrigerant oil until you can see the oil level.

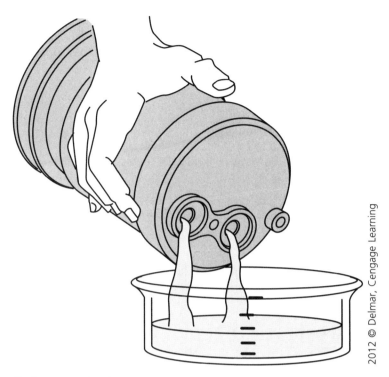

2012 © Delmar, Cengage Learning

16. Referring to the figure above, Technician A says that the old compressor oil should be drained and measured when replacing an A/C compressor. Technician B says that the old oil should be added to the new compressor. Who is correct?

    A. A only
    B. B only
    C. Both A and B
    D. Neither A nor B

17. A cracked A/C compressor mounting plate could cause all of the following symptoms EXCEPT:

    A. Drive belt wear.
    B. Internal compressor damage.
    C. Vibration with the A/C compressor clutch engaged.
    D. Drive belt squeal or chatter.

18. Technician A says that one ounce of refrigerant oil should be added to the accumulator/drier prior to installing it on a truck. Technician B says that five ounces of oil should be added to the suction line prior to installing it on a truck. Who is correct?

    A. A only
    B. B only
    C. Both A and B
    D. Neither A nor B

19. Which of the following should be used to lubricate replacement A/C hose o-rings?

    A. Petroleum jelly
    B. Transmission fluid
    C. Silicone grease
    D. Refrigerant oil

20. Technician A says that deformed or improperly aligned condenser mounting insulators will damage the A/C compressor. Technician B says that deformed or improperly aligned condenser mounting insulators could damage the condenser and refrigerant lines. Who is correct?

    A. A only
    B. B only
    C. Both A and B
    D. Neither A nor B

21. The thermal bulb on an expansion valve must be installed in contact with:

    A. The evaporator outlet tube.
    B. The condenser fins.
    C. The suction hose.
    D. The refrigerant.

22. A whistling noise coming from under the passenger side dash with the blower motor on high speed might indicate:

    A. A clogged evaporator drain.
    B. A cracked evaporator case.
    C. A broken blend door cable.
    D. A low refrigerant charge.

23. A heater does not supply the cab with enough heat. The coolant level and blower test OK. Technician A says an improperly adjusted temperature control cable could be the cause. Technician B says a clogged heater core could be the cause. Who is correct?

    A. A only
    B. B only
    C. Both A and B
    D. Neither A nor B

24. Coolant conditioner performs all of the following tasks EXCEPT:

    A. It raises the boiling point of the coolant.
    B. It inhibits rust and debris from the coolant.
    C. It lubricates the cooling system internally.
    D. It prevents cavitation corrosion of wet cylinder liners.

25. A coiled spring inside a radiator hose is used to:

    A. Pre-form the hose.
    B. Prevent the hose from collapsing.
    C. Increase the hose burst pressure.
    D. Eliminate cavitation.

26. A cooling system pressure tester can be used to test:

    A. Thermostats.
    B. Radiators, pressure caps, and hoses.
    C. A/C leaks.
    D. The blend door actuator diaphragm.

27. The LEAST LIKELY cause of poor coolant circulation in a truck with a down-flow radiator is:

    A. A defective thermostat.

    B. An eroded water pump impeller.

    C. A collapsed upper radiator hose.

    D. A collapsed lower radiator hose.

28. All of the following statements about the chassis air system are true EXCEPT:

    A. It is important to keep water drained from the system.

    B. System air can be used to operate the fan clutch.

    C. System air can be used to operate the radiator shutters.

    D. System air actuates the engine thermostat.

29. A blown HVAC system fuse could indicate any of the following EXCEPT:

    A. A short circuit to ground in the blower circuit.

    B. A short circuit to ground in the blend door actuator.

    C. A short circuit in the engine coolant temperature sensor (ECT).

    D. A damaged wiring harness connector.

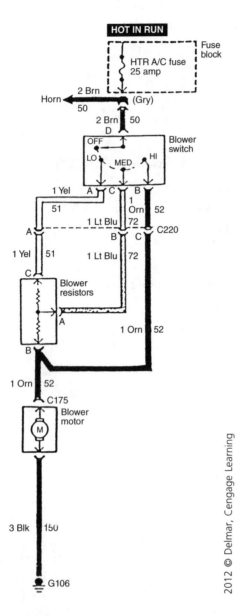

30. Refer to the figure above. The blower motor works in LO and HI positions but does not work on MED. Technician A says the problem could be an open blower resistor. Technician B says the problem could be the blower motor switch. Who is correct?

A. A only

B. B only

C. Both A and B

D. Neither A nor B

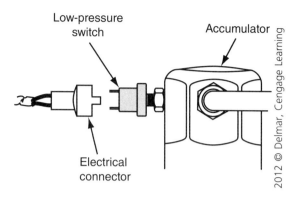

31. All of the following conditions would cause the contacts of the switch in the figure to be "open" EXCEPT:

     A. Cold temperature in the evaporator core.

     B. An inoperative condenser fan.

     C. An empty refrigerant system.

     D. Ambient temperature of 10° F (–12.2° C).

32. An electric cooling fan motor can be controlled by any of the following EXCEPT:

     A. An electronic relay.

     B. An independent electronic module.

     C. A multiplexed electronic module.

     D. An air solenoid controller.

33. Before replacing an electric blend air door actuator, the technician should:

     A. Ensure that the batteries are removed from the vehicle.

     B. Ensure that the batteries have a good ground.

     C. Ensure that the blend door moves freely.

     D. Ground himself to the vehicle.

34. A truck that uses an HVAC system with a vacuum control panel produces no cool air out of the dash outlets. The air only discharges out of the defrost outlets. Which of these items is the most likely cause?

     A. A leaking dash vacuum switch

     B. A defective A/C compressor

     C. Loss of vacuum to the control panel

     D. A defective heater control valve

35. All of the following are examples of HVAC control systems types EXCEPT:

     A. Orifice tube climate control.

     B. Manual climate control.

     C. Semi-automatic climate control.

     D. Automatic climate control.

36. A technician has just recovered the refrigerant from an A/C system. Technician A says that the waste oil bottle should be checked to see how much oil was removed during the recovery process. Technician B says that the recovery weight scale should be checked to see how much refrigerant was removed during the process. Who is correct?

    A. A only
    B. B only
    C. Both A and B
    D. Neither A nor B

37. Technician A says that according to new environmental laws, shut-off valves must be placed in the closed position every time the A/C system is turned off. Technician B says that according to new environmental laws, shut-off valves must be located no more than 12 inches from test hose service ends. Who is correct?

    A. A only
    B. B only
    C. Both A and B
    D. Neither A nor B

38. A refrigerant identifier is connected to a truck A/C system and gives the reading of 90 percent R-134a and 10 percent R-12. Technician A says that this system can be safely recovered into the R-134a recovery machine. Technician B says that this system has likely been retrofitted using non-standard procedures. Who is correct?

    A. A only
    B. B only
    C. Both A and B
    D. Neither A nor B

39. Which of these terms correctly describes refrigerant which has been removed from a system and stored in an external container?

    A. Recycled
    B. Recovered
    C. Reclaimed
    D. Refined

40. The test for non-condensable gases in recovered/recycled refrigerant involves:

    A. Comparing the pressure of the recovered refrigerant in the container to the theoretical pressure of pure refrigerant at a given temperature.
    B. Comparing the atmospheric pressure to the relative humidity.
    C. Comparing the container pressure with the size of the container.
    D. Testing the refrigerant with a halogen leak detector.

PREPARATION EXAM 5

1. Which of the following methods would be the LEAST LIKELY way to verify the driver's complaint about an HVAC problem on a truck?

    A. Operate the system in question in the service bay.

    B. Perform a thorough inspection.

    C. Read the repair order.

    D. Perform a road test.

2. An automatic temperature control (ATC) system is being diagnosed. Technician A says that this system must use an electronic control head. Technician B says that this system can be diagnosed with a scan tool. Who is correct?

    A. A only

    B. B only

    C. Both A and B

    D. Neither A nor B

3. A compressor has excessive air gap at the clutch drive plate. Technician A says that the clutch could squeal when the compressor engages. Technician B says that the clutch could squeal when the engine is quickly accelerated while running the air conditioning. Who is correct?

    A. A only

    B. B only

    C. Both A and B

    D. Neither A nor B

4. Which of the following practices is LEAST LIKELY to be performed during a heater core replacement?

    A. Inspect the blend and mode doors after reassembling the duct box to assure correct operation.

    B. Completely drain the coolant from the radiator and engine.

    C. Disassemble the instrument panel to gain access to the heater core.

    D. Install hose crimping pliers on the heater hoses.

5. An HVAC system outputs air at a constant temperature, regardless of the temperature setting. What is the most likely cause for this condition?

    A. Low refrigerant charge

    B. Low coolant level

    C. Compressor clutch failure

    D. Broken blend door cable

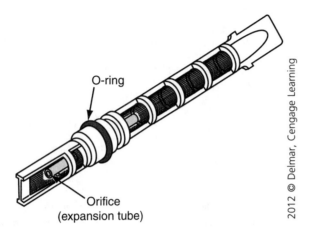

O-ring

Orifice
(expansion tube)

2012 © Delmar, Cengage Learning

6. Referring to the figure, Technician A says that the device is an orifice tube and is used on truck systems that use a receiver/drier. Technician B says that the device meters refrigerant into the evaporator core. Who is correct?

   A. A only
   B. B only
   C. Both A and B
   D. Neither A nor B

7. Which of these characteristics does the R-134a refrigerant possess?

   A. No odor
   B. Faint ether-like odor
   C. Strong rotten egg odor
   D. Cabbage-like odor

8. Which of the following procedures would be LEAST LIKELY used to determine the freeze protection level of the engine coolant?

   A. Test strips
   B. Visual inspection
   C. Hydrometer
   D. Refractometer

9. An A/C system has been open to the atmosphere during a repair. What is the minimum recommended length of time that the vacuum pump should be operated for evacuation?

   A. 20 minutes
   B. 10 minutes
   C. 15 minutes
   D. 30 minutes

10. Technician A says the A/C system can be charged through the low side with the system running. Technician B says inverting the refrigerant container causes low-pressure refrigerant vapor to be charged into the system. Who is correct?

   A. A only
   B. B only
   C. Both A and B
   D. Neither A nor B

11. All of the following electronic devices could be used to control the operation on truck A/C systems EXCEPT:

    A. Pressure cycling switch.
    B. Engine oil temperature sensor.
    C. Evaporator temperature switch.
    D. Binary pressure switch.

12. Technician A says that some HVAC electrical control panels have replaceable bulbs. Technician B says that the HVAC electrical control panel can be replaced without removing the instrument panel trim bezel. Who is correct?

    A. A only
    B. B only
    C. Both A and B
    D. Neither A nor B

13. The A/C compressor clutch will not engage in any mode. The clutch engages when a technician installs a jumper wire across the terminals of the low-pressure cut-out switch connector. Technician A says that the low-pressure cut-out switch must be defective. Technician B says that the refrigerant charge could be low. Who is correct?

    A. A only
    B. B only
    C. Both A and B
    D. Neither A nor B

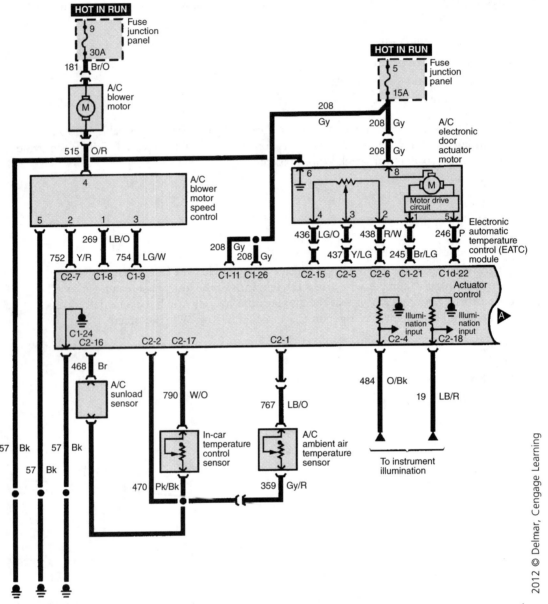

14. Referring to the figure, when measuring the voltage drop in the A/C computer ground as shown, the technician connects a voltmeter from computer terminal C1-24 internal ground to the external ground. With the ignition on, the maximum allowable voltage drop should be which of the following?

A. 0.05 volts

B. 0.3 volts

C. 0.5 volts

D. 0.8 volts

15. Technician A says that you can test the A/C compressor clutch coil with an ohmmeter. Technician B says that connecting battery power to one terminal of the coil and grounding the other terminal can test the A/C compressor clutch coil. Who is correct?

A. A only

B. B only

C. Both A and B

D. Neither A nor B

16. Which of the following statements is true about checking the A/C compressor lubricant level?
    A. The A/C system must first be evacuated for at least 30 minutes.
    B. The compressor must not be operated for at least 24 hours before checking the lubricant level.
    C. The oil level should be up to the full mark on the dipstick.
    D. The old lubricant must be measured before the new lubricant is added.

17. Technician A says that the compressor reed valves can often be replaced without discharging the A/C system. Technician B says that damaged compressor reed valves can send debris into the condenser. Who is correct?
    A. A only
    B. B only
    C. Both A and B
    D. Neither A nor B

18. All of the following statements about a computer-controlled A/C system are correct EXCEPT:
    A. Some actuator motors are calibrated automatically in the self-diagnostic mode.
    B. A/C diagnostic trouble codes (DTC) represent the exact fault in a specific component.
    C. The actuator control rods must be calibrated manually on some systems.
    D. The actuator motor control rods should only require adjustment after motor replacement or adjustment.

19. Airflow through the A/C condenser is significantly affected by all of the following EXCEPT:
    A. Debris trapped in the fins.
    B. Relative humidity of the outside air.
    C. Bent cooling fins.
    D. Vehicle speed.

20. Technician A says that the heater control valve can be controlled by oil pressure. Technician B says that the heater control valve can be cable operated. Who is correct?
    A. A only
    B. B only
    C. Both A and B
    D. Neither A nor B

21. A screen is located in the orifice tube of an A/C system. Technician A says that the screen is a filter used to prevent particulate from circulating through the system. Technician B says that the screen is used to improve atomization of the refrigerant. Who is correct?
    A. A only
    B. B only
    C. Both A and B
    D. Neither A nor B

22. The inside of a truck windshield has an oily film and the A/C cooling is poor. Technician A says a plugged HVAC evaporator drain may cause this oil film. Technician B says this film may be caused by a large leak in the evaporator core. Who is correct?
    A. A only
    B. B only
    C. Both A and B
    D. Neither A nor B

23. A truck is being diagnosed for an A/C system problem. The operating pressures are checked. The results show the low-side pressure is too high and the high-side pressure is too low. Technician A says that the compressor may have a faulty reed valve. Technician B says that an overcharge of refrigerant oil is a possible cause. Who is correct?

    A. A only
    B. B only
    C. Both A and B
    D. Neither A nor B

24. Which of the following is the expected result from raising the pressure in the cooling system?

    A. Lower coolant boiling point
    B. Elevated coolant boiling point
    C. Corrosion prevention in the cooling system
    D. No change in the coolant boiling point

25. All of the following are good methods of verifying that an engine thermostat opens EXCEPT:

    A. Feeling the upper radiator hose.
    B. Watching the temperature gauge.
    C. Watching for motion in the upper radiator tank.
    D. Watching the surge tank.

26. A cooling fan can be controlled by any of the following EXCEPT:

    A. An air clutch.
    B. A thermostatic spring and fan clutch fluid.
    C. An electrically actuated clutch.
    D. A hydraulic switch.

27. When an expansion valve is properly installed, where should the thermal bulb and capillary tube be positioned?

    A. Fastened to the condenser fins using epoxy
    B. Located in the accumulator
    C. Held in contact with the evaporator outlet using insulating tape
    D. As an integral part of the orifice tube assembly

28. What could a faint ether-like odor coming from the panel vents in the NORMAL A/C mode indicate?

    A. The evaporator core is leaking R-134a.
    B. The evaporator core is leaking R-12.
    C. The cold starting system is malfunctioning.
    D. The heater core is leaking.

29. A truck is being diagnosed for having three compressor failures in a six-month period. Each compressor has failed in the area of the front drive clutch overheating. Technician A says that a weak battery pack could be causing this problem. Technician B says that a faulty A/C clutch relay could be causing the problem. Who is correct?

    A. A only
    B. B only
    C. Both A and B
    D. Neither A nor B

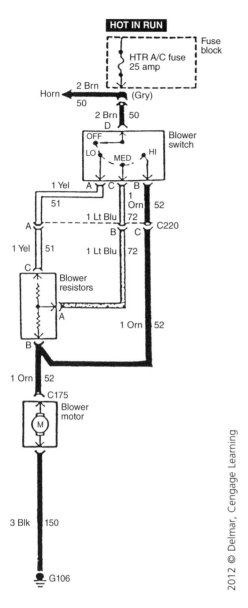

30. Referring to the figure above, Technician A says that the blower motor has four speeds. Technician B says that the blower resistor has two resistors. Who is correct?

   A. A only

   B. B only

   C. Both A and B

   D. Neither A nor B

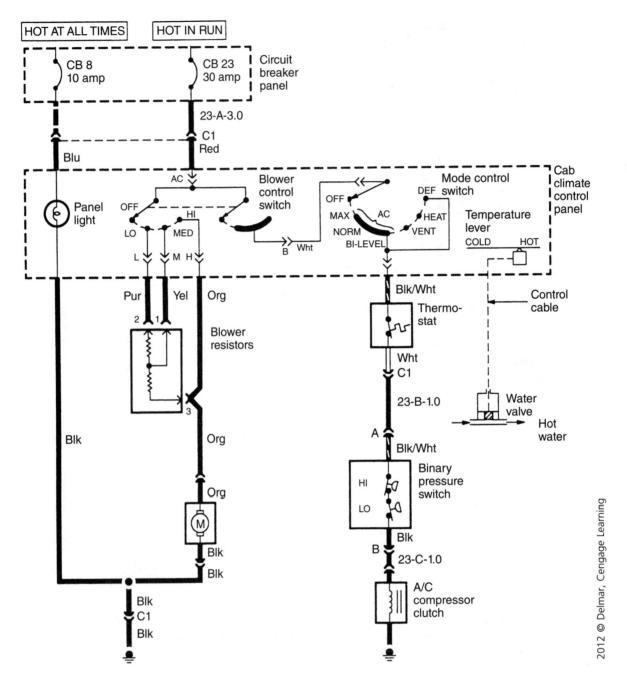

31. Referring to the figure above, the compressor clutch is inoperative in the DEF mode but operates properly in all other A/C modes. Technician A says the low side of the binary switch may have an open circuit. Technician B says the defrost switch contacts may have an open circuit. Who is correct?

    A. A only

    B. B only

    C. Both A and B

    D. Neither A nor B

32. Technician A says that the evaporator must be removed from the vehicle if the presence of evaporator lubricant is to be checked. Technician B says that the refrigerant oil is distributed throughout the A/C system. Who is correct?

    A. A only
    B. B only
    C. Both A and B
    D. Neither A nor B

33. Which of the following electronic devices would be LEAST LIKELY to control the operation of a truck A/C system?

    A. Low-pressure switch
    B. Dual pressure switch
    C. A/C pressure sensor
    D. Fuel temperature sensor

34. Which of the following procedures would be the most likely method of repairing an air leak in the HVAC control panel?

    A. Replacement of the control panel
    B. Replacement of the pintle o-rings
    C. Repacking the selector body with grease
    D. Replacing the selector levers

35. A truck that has an automatic temperature control (ATC) system is being diagnosed. Technician A says that these systems have a self-diagnostic system that sets trouble codes when certain problems occur. Technician B says that ATC trouble codes can be retrieved with a scan tool. Who is correct?

    A. A only
    B. B only
    C. Both A and B
    D. Neither A nor B

36. A heavy truck with a faulty A/C pressure sensor is being diagnosed. Technician A says that this device is a three-wire feedback sensor that is used to signal A/C system pressure to a control module. Technician B says that this device is typically mounted in the low side of the A/C system. Who is correct?

    A. A only
    B. B only
    C. Both A and B
    D. Neither A nor B

37. Technician A says that all A/C repair shops are required to use SAE-approved recovery equipment. Technician B says that all individuals who service A/C systems must be certified by a recognized body on how to properly handle refrigerants. Who is correct?

    A. A only
    B. B only
    C. Both A and B
    D. Neither A nor B

38. Federal laws require safe refrigerant handling procedures. Technician A says that stored refrigerant should be kept warm by providing a heat source near the storage area. Technician B says that recovered refrigerant should be kept in a DOT 39 cylinder. Who is correct?

    A. A only
    B. B only
    C. Both A and B
    D. Neither A nor B

39. All of the following steps should be performed after an HVAC repair has been made EXCEPT:

    A. A thorough visual inspection.
    B. Clearing diagnostic codes.
    C. Operating the system to check performance.
    D. Recovering the refrigerant.

40. Technician A says that a failed water pump bearing produces a metallic knock at the rear of the engine. Technician B says that a failed compressor pulley bearing will produce noise when the compressor is not engaged. Who is correct?

    A. A only
    B. B only
    C. Both A and B
    D. Neither A nor B

PREPARATION EXAM 6

1. Which of the following examples would be a benefit of reviewing the past maintenance documents for a truck?

    A. Determining how much money the owner has spent on service
    B. Finding out how to perform the road test
    C. Assisting in performing the visual inspection
    D. Determining if the truck has had past HVAC-related repairs

2. A truck that has an automatic temperature control (ATC) system is being diagnosed. Technician A says that ATC trouble codes are retrieved by depressing a sequence of buttons on the ATC control head. Technician B says that the trouble code reveals exactly what needs to be repaired in an ATC system. Who is correct?

    A. A only
    B. B only
    C. Both A and B
    D. Neither A nor B

3. Technician A says that the line exiting the condenser should be hotter than the line entering the condenser. Technician B says that the suction line should be cold to the touch when the A/C system is operating. Who is correct?

    A. A only
    B. B only
    C. Both A and B
    D. Neither A nor B

4. A truck technician discovers a heater core leak that requires its replacement. Technician A says that the entire cooling system will need to be drained prior to removal of the heater core. Technician B says that the replacement heater core should be installed without insulation tape. Who is correct?

    A. A only
    B. B only
    C. Both A and B
    D. Neither A nor B

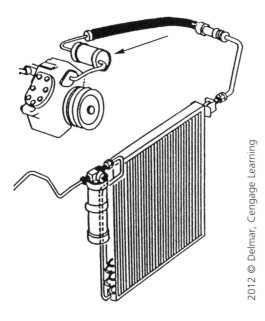

2012 © Delmar, Cengage Learning

5. Referring to the figure above, Technician A says the device indicated by the arrow is used on some trucks to reduce compressor noise. Technician B says that the device indicated by the arrow can be flushed out if it gets restricted. Who is correct?

    A. A only
    B. B only
    C. Both A and B
    D. Neither A nor B

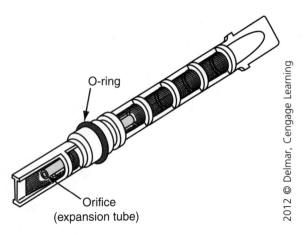

O-ring

Orifice
(expansion tube)

2012 © Delmar, Cengage Learning

6. What is the most likely drying component to be used with the metering device in the figure above?

    A. Valves in receiver/drier
    B. Accumulator/drier
    C. Muffler drier
    D. Receiver/drier

7. Technician A says that the easiest way to identify the type of refrigerant that a system should use is to observe the service port fittings. Technician B says that R-134a is the only refrigerant that may be vented to the atmosphere. Who is correct?

    A. A only
    B. B only
    C. Both A and B
    D. Neither A nor B

8. When a technician pressure tests a cooling system, there are no obvious external leaks but the system cannot maintain pressure. Which of the following is the most likely cause of this problem?

    A. Leaking evaporator
    B. Defective heater valve
    C. Stuck open thermostat
    D. Blown head gasket in the engine

9. Which of the following items is most likely to cause elevated high-side pressure in an A/C system?

    A. Restricted airflow through condenser
    B. Stuck open thermostat
    C. Leaking thermal bulb
    D. Inoperative compressor

10. Where would you position the leak detector sensor to detect a refrigerant leak?

    A. Within three inches of the fitting
    B. Just below the fitting
    C. Right next to the fitting
    D. Just above the fitting

11. Technician A says that moisture can be removed from the A/C system after charging the system with new refrigerant. Technician B says moisture that enters the A/C circuit will be harmful to the system and cause poor performance. Who is correct?

    A. A only
    B. B only
    C. Both A and B
    D. Neither A nor B

12. Which of the following functions would be LEAST LIKELY to be performed with a scan tool on a climate control computer?

    A. Display sensor data.
    B. Calibrate the air handling door actuators.
    C. Reprogram the blower motor.
    D. Display switch data.

13. Which of the following components would be LEAST LIKELY used as a de-icing device in the A/C system?

    A. Variable displacement compressor
    B. Pressure cycling switch
    C. Evaporator temperature sensor
    D. Pressure release valve

14. A serpentine belt is being replaced on a diesel-powered truck and the tensioner will not snap back after being released. Technician A says that the tensioner needs to be greased to get it to function correctly. Technician B says that the idler pulley is jammed and will need to be replaced. Who is correct?

    A. A only
    B. B only
    C. Both A and B
    D. Neither A nor B

15. Which of the following A/C gauge set readings would most likely indicate the A/C compressor reed valves are worn out?

    A. High-side pressure too low and low-side pressure too high
    B. No pressure on either gauge
    C. Pressure readings that are too low for both gauges
    D. Pressure readings that are too high for both gauges

16. Technician A says that when installing a new or remanufactured compressor you should first turn it over by hand and drain any oil shipped with the compressor, and then install the correct amount of the specified oil before installing it. Technician B says that some compressors are shipped without oil. Who is correct?

    A. A only
    B. B only
    C. Both A and B
    D. Neither A nor B

17. A knocking noise is heard in the compressor area that is audible when the compressor is engaged, but that goes away when the compressor turns off. Technician A says that loose compressor mounting bolts could be the cause. Technician B says that a discharge line rubbing a compressor mounting bracket could be the cause. Who is correct?

    A. A only
    B. B only
    C. Both A and B
    D. Neither A nor B

18. A routine A/C maintenance service should include all of the following EXCEPT:

    A. Tightening the condenser lines.
    B. Removing debris from the condenser fins.
    C. Straightening the condenser fins.
    D. Checking the condenser mounts.

19. All of the following features could be found on a truck with an automatic climate control system EXCEPT:

    A. Cabin temperature sensor.
    B. Electronic climate control head.
    C. HVAC logic device.
    D. Blower resistor.

20. All of the following statements about coolant control valves are true EXCEPT:

    A. It controls the flow of coolant through the heater core.
    B. It is part of the water pump.
    C. It may be cable-operated.
    D. It may be vacuum-operated.

21. Which of the following is the location for the expansion valve?

    A. Inlet line of the evaporator
    B. Outlet line of the evaporator
    C. Inlet line of the compressor
    D. Outlet line of the condenser

22. The A/C compressor high-pressure relief valve:

    A. Is calibrated by shimming it to the proper depth.
    B. Must be replaced if it ever vents refrigerant from the system.
    C. Will reset itself when A/C system pressure returns to a safe level.
    D. Is not used in R-134a systems.

23. Which of the following features is LEAST LIKELY to be found on a truck with an automatic climate control?

    A. Cable-operated mode door actuator
    B. Cabin temperature sensor
    C. Electronic climate control head
    D. Electronic blend door actuator

24. The LEAST LIKELY cause of poor coolant circulation in a truck with a down-flow radiator is:

    A. A defective thermostat.

    B. An eroded water pump impeller.

    C. A collapsed upper radiator hose.

    D. A collapsed lower radiator hose.

25. All of these statements about cooling system service are true EXCEPT:

    A. When the cooling system pressure is increased, the boiling point is decreased.

    B. If more antifreeze is added to the coolant mix, the boiling point is increased.

    C. A good quality ethylene glycol antifreeze contains a corrosion inhibitor.

    D. Coolant solutions must be recovered, recycled, or handled as hazardous material.

26. Technician A says that a cracked fan blade should be welded. Technician B says that a cracked fan blade can be repaired with epoxy. Who is correct?

    A. A only

    B. B only

    C. Both A and B

    D. Neither A nor B

27. Technician A says that you should replace the receiver/drier if the sight glass appears cloudy because it indicates a ruptured desiccant pack. Technician B says that you should only replace the receiver/drier if it has a leak. Who is correct?

    A. A only

    B. B only

    C. Both A and B

    D. Neither A nor B

28. Which of the following statements is the LEAST LIKELY result of a touch test performed on a normal A/C system?

    A. The compressor discharge line is hot to the touch.

    B. The line exiting the orifice tube is cold with a frost ring around it.

    C. The suction line is cold with condensation droplets on it.

    D. The line exiting the condenser is not as hot as the line entering the condenser.

29. What does the pressure cycling switch located on the accumulator sense?

    A. Outside temperature

    B. Accumulator pressure

    C. Accumulator temperature

    D. Engine compartment temperature

30. A coolant temperature sensor is classified as negative temperature coefficient (NTC). Technician A says this is a thermistor in which internal resistance decreases in proportion to temperature rise. Technician B says that in most truck engine cooling systems, a thermistor is supplied with battery voltage (V-Bat) and returns a portion of it as a signal. Who is correct?

    A. A only
    B. B only
    C. Both A and B
    D. Neither A nor B

31. A truck is being diagnosed for a problem with the blend door actuator motor. The motor runs when the temperature setting is changed, but the blend door does not move. Which of the following conditions would be the most likely cause of this problem?

    A. Defective control module
    B. Defective actuator feedback device
    C. Defective drive gear in the actuator
    D. Improperly adjusted mode door linkage

32. All of the statements about replacing an HVAC control panel are true EXCEPT:

    A. The negative battery cable should be removed before servicing the control panel.
    B. The refrigerant must be recovered before removing the control panel.
    C. If the truck contains a supplemental restraint system (SRS), a technician must wait the specified period after removing the negative battery cable.
    D. Self-diagnostic tests may indicate a defective control panel in an automatic temperature control (ATC) system.

33. Technician A says a container of PAG refrigerant oil must be kept closed when not in use to prevent the oil from absorbing moisture. Technician B says that oil must be added to all new components during installation. Who is correct?

    A. A only
    B. B only
    C. Both A and B
    D. Neither A nor B

34. All of the following statements about air-controlled HVAC systems in heavy-duty trucks are true EXCEPT:

    A. Air cylinders are used to open shutters.
    B. Coolant control valves can be controlled using chassis air.
    C. Air cylinders are used to control mode and blend air doors.
    D. Air leaks may cause mode doors to move slowly or to be totally inoperative.

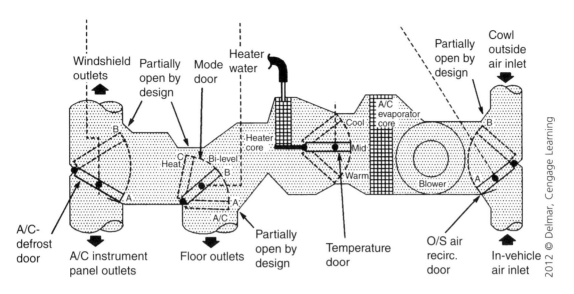

35. Referring to the figure above, the outside air recirculation door is stuck in position A. Technician A says under this condition outside air is drawn into the HVAC inlet. Technician B says under this condition in-vehicle air is recirculated. Who is correct?

    A. A only

    B. B only

    C. Both A and B

    D. Neither A nor B

36. The best instrument to use when troubleshooting an A/C electronic circuit with solid-state components is:

    A. Digital multi-meter (DMM).

    B. Self-powered test lamp.

    C. Analog volt/ohmmeter (VOM).

    D. 12V test lamp.

37. A refrigerant identifier is connected to a truck and gives readings of 100 percent R-134a and 0.0 percent R-12. Technician A says that this system can be safely recovered into the R-134a recovery machine. Technician B says that refrigerant identifiers should be used on every vehicle prior to connecting any A/C equipment. Who is correct?

    A. A only

    B. B only

    C. Both A and B

    D. Neither A nor B

| Pressure-Temperature Relationship | | | | |
Temperature °F (°C)	R-12 PSIG (bar/kg/cm2)		R-134A PSIG (bar/kg/cm2)	
–15 (–26.1)	.2.5	(.17/.18)	0	(0)
–10 (–23.3)	.4.5	(.31/.32)	2.0	(.14/.14)
–5 (–20.5)	.6.7	(.46/.03)	4.1	(.28/.29)
0 (–17.8)	.9.2	(.63/.65)	6.5	(.45/.46)
5 (–15.0)	.11.8	(.81/.83)	9.1	(.63/.64)
10 (–12.2)	.14.7	(1.0/1.0)	12.0	(.89/.84)
15 (–9.4)	.17.7	(1.2/1.2)	15.1	(1.0/1.2)
20 (–6.7)	.21.1	(1.5/1.5)	18.4	(1.3/1.3)
25 (–3.9)	.24.6	(1.7/1.7)	22.1	(1.5/1.6)
30 (–1.1)	.28.5	(2.0./2.0)	26.1	(1.8/1.8)
35 (1.7)	.32.6	(2.2/2.3)	30.4	(2.1/2.1)
40 (4.4)	.37.0	(2.6/2.6)	35.0	(2.4/2.5)
45 (7.2)	.41.7	(2.9/3.0)	40.0	(2.8/2.8)
50 (10.0)	.46.7	(3.2/3.3)	45.4	(3.1/3.2)
55 (12.8)	.52.1	(3.6/3.7)	51.2	(3.5/3.6)
60 (15.6)	.57.8	(4.0/4.1)	57.4	(4.0/4.0)
65 (18.3)	.63.8	(4.4/4.5)	64.0	(4.4/4.5)
70 (21.1)	.70.2	(4.8/5.0)	71.1	(5.0/5.0)
75 (23.9)	.77.0	(5.3/5.4)	78.6	(5.4/5.5)
80 (26.7)	.84.2	(5.8/6.0)	86.7	(6.0/6.1)
85 (29.4)	.91.7	(6.3/6.4)	95.2	(6.6/6.7)
90 (32.2)	.99.7	6.9/7.0	104.3	(7.2/7.3)
95 (35.0)	.108.2	(7.5/7.6)	113.9	(7.9/8.0)
100 (37.8)	.117.0	(8.1/8.2	124.1	(8.6/8.7)
105 (40.6)	.126.4	(8.7/8.9)	134.9	(9.3/9.5)
110 (43.3)	.136.2	(9.4/9.6)	146.3	(10.1/10.3)
115 (46.1)	.146.5	(10.1/10.3)	158.4	(11.0/11.1)
120 (48.9)	.157.3	(11.0/11.1)	171.1	(11.8/12.0)

2012 © Delmar, Cengage Learning

38. Referring to the table above, a storage container or A/C system containing R-134a (at rest) and subject to an ambient temperature of 70° F (21° C) will have an internal gauge pressure of approximately:

   A. 220 psi (1517 kPa).
   B. 125 psi (862 kPa).
   C. 71 psi (490 kPa).
   D. 30 psi (207 kPa).

39. Which of the following repair processes is the best description of the term recovery?

   A. The process of pulling the A/C system into a vacuum in order to remove the air and moisture from the system.
   B. The process of removing and weighing the refrigerant from an A/C system.
   C. The process of filtering the refrigerant to remove impurities from it.
   D. The process of adding refrigerant to an A/C system after repair work has been performed.

40. Which of the following problems would most likely cause a loud hissing sound accompanied by a momentary release of refrigerant vapor under the hood?

   A. Blocked thermal expansion valve
   B. Condenser coated with mud and debris
   C. Restricted evaporator core
   D. Faulty compressor

Answer Keys and Explanations

Included in this section are the answer keys for each preparation exam, followed by individual, detailed answer explanations and a reference identifying the designated task area being assessed by each specific question. This additional reference information may prove useful if you need to refer back to the task list located in Section 4 of this book for additional support.

PREPARATION EXAM 1—ANSWER KEY

1.	C	21.	A
2.	D	22.	A
3.	C	23.	C
4.	D	24.	C
5.	A	25.	B
6.	C	26.	C
7.	A	27.	D
8.	D	28.	C
9.	D	29.	C
10.	D	30.	B
11.	C	31.	C
12.	A	32.	B
13.	D	33.	B
14.	C	34.	D
15.	C	35.	A
16.	A	36.	D
17.	A	37.	C
18.	B	38.	C
19.	B	39.	C
20.	A	40.	B

PREPARATION EXAM 1—EXPLANATIONS

TASK A.1

1. Technician A says that reviewing past maintenance records will help determine if the truck has had past HVAC-related repairs. Technician B says that it is necessary to perform a road test to verify the problem on some trucks. Who is correct?

   A. A only

   B. B only

   C. Both A and B

   D. Neither A nor B

   Answer A is incorrect. Technician B is also correct.

   Answer B is incorrect. Technician A is also correct.

   Answer C is correct. Both Technicians are correct. Reviewing past maintenance records would assist the technician by revealing the past HVAC-related repairs on the truck. Some driver complaints require the technician to perform a road test in order to verify the problem.

   Answer D is incorrect. Both Technicians are correct.

TASK A.2

2. A faint hissing noise is heard from the area of the evaporator immediately after shutting down the engine with the compressor clutch engaged. Technician A says that the A/C system has a leak. Technician B says that the thermal expansion valve (TXV) is defective. Who is correct?

   A. A only

   B. B only

   C. Both A and B

   D. Neither A nor B

   Answer A is incorrect. A faint hissing noise after shutting down the engine is caused by the equalization of refrigerant pressure in the A/C system.

   Answer B is incorrect. A defective TXV would cause poor cooling accompanied by irregular pressures.

   Answer C is incorrect. Neither Technician is correct.

   Answer D is correct. Neither Technician is correct. It is normal to have an audible noise in the A/C system after shutting the engine off. The noise comes from the equalizing pressures from each side of the system.

TASKS A.2, B.2.3

3. Technician A says that a weak belt tensioner can cause the drive belt to squeal momentarily as the A/C compressor engages. Technician B says that a weak belt tensioner can cause the drive belt to squeal when the engine is quickly accelerated. Who is correct?

   A. A only

   B. B only

   C. Both A and B

   D. Neither A nor B

   Answer A is incorrect. Technician B is also correct.

   Answer B is incorrect. Technician A is also correct.

   Answer C is correct. Both Technicians are correct. A weak belt tensioner could cause the drive belt to squeal when the A/C clutch engages, as well as when the engine is quickly accelerated, due to increased load at these times.

   Answer D is incorrect. Both Technicians are correct.

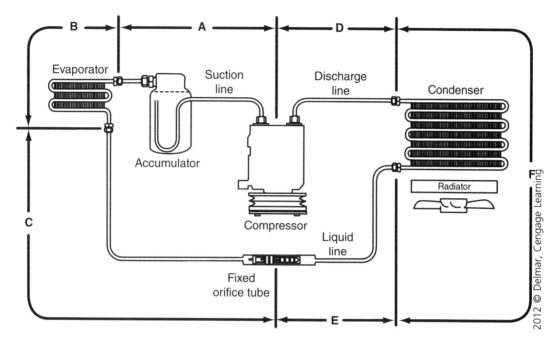

4. Referring to the figure above, which of the following conditions would most likely cause ice to form in the section labeled "F"?

   A. Air in the refrigerant system
   B. Inoperative compressor
   C. Faulty condenser fan
   D. Internal blockage in the condenser

TASK A.3

Answer A is incorrect. Air in the refrigerant system would cause elevated high-side pressures and reduced A/C system performance, but it would not cause ice to form in the high side of the system.

Answer B is incorrect. An inoperative compressor would cause the A/C system to have equal pressures on both sides of the system, but it would not cause ice to form in the high side of the system.

Answer C is incorrect. A faulty condenser fan could cause elevated high-side pressures and reduced A/C system performance when the truck is driven at low speeds or when it is stopped.

Answer D is correct. An internal blockage in the condenser could cause ice to form in the high side of the A/C system. The high side of the A/C system should always be warmer than the low side of the system.

5. The refrigerant line leading from the evaporator to the compressor contains refrigerant as a:

   A. Low-pressure gas.
   B. Low-pressure liquid.
   C. High-pressure gas.
   D. High-pressure liquid.

TASK A.4

Answer A is correct. Refrigerant returns to the compressor as a low-pressure gas. This line is known as the suction line and is very cold on a normally operating system.

Answer B is incorrect. Refrigerant in an operating A/C system is a low-pressure liquid after it passes through the metering device. While in the evaporator, the refrigerant changes from a low-pressure liquid into a low-pressure gas.

Answer C is incorrect. Refrigerant leaves the compressor as a high-pressure gas.

Answer D is incorrect. Refrigerant in the liquid line is a high-pressure liquid. The liquid line connects the condenser to the evaporator. The condenser changes high-pressure vapor into a high-pressure liquid by allowing the refrigerant to give up heat and condense into a liquid.

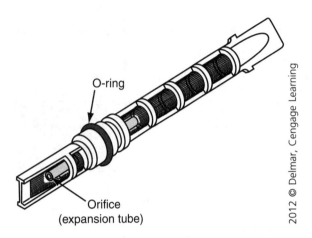

O-ring

Orifice
(expansion tube)

2012 © Delmar, Cengage Learning

TASK A.4

6. Referring to the figure above, what function does the device shown perform for the refrigerant system?

   A. Meters refrigerant into the compressor

   B. Filters refrigerant into the condenser

   C. Meters refrigerant into the evaporator core

   D. Filters refrigerant into the compressor

   Answer A is incorrect. The suction line routes refrigerant into the compressor.

   Answer B is incorrect. The discharge line routes refrigerant into the condenser. A filter is not typically located at this part of the system.

   Answer C is correct. The orifice tube meters high-pressure liquid refrigerant into the evaporator core. After the refrigerant passes the orifice tube, it becomes a low-pressure atomized liquid.

   Answer D is incorrect. The suction line routes refrigerant into the compressor. Some systems use a screen on the compressor inlet connection that prevents debris from entering the compressor.

TASK B.1.1

7. A driver complains that he is unable to change the output temperature of his HVAC system. What is the most likely cause of this problem?

   A. Broken blend door cable

   B. Defective compressor clutch

   C. Clogged orifice tube

   D. Defective blower switch

   Answer A is correct. A broken blend door cable will make HVAC output temperature uncontrollable. The blend door controls the temperature of the air by routing the air through or around the heater core. When the control is set to cold, the blend door blocks the air from going past the heater core. When the control is set for hot, the blend door routes all of the air past the heater core. When the control is set between hot and cold, the blend mixes the air by allowing some of the air to go past the heater core and some to by-pass it.

   Answer B is incorrect. A defective compressor clutch will not affect heater temperature control.

   Answer C is incorrect. A clogged orifice tube will not affect heater temperature control.

   Answer D is incorrect. An inoperative blower motor has no effect on temperature control.

8. A/C system pressures vary with all of the following EXCEPT:

   A. Altitude.

   B. Ambient temperature.

   C. Ambient air humidity.

   D. Cab humidity.

TASK B.1.3

   Answer A is incorrect. A/C system pressures will vary with altitude changes.

   Answer B is incorrect. A/C system pressures will vary with ambient temperature. Higher ambient temperature causes the pressures to rise, especially the high-side pressure.

   Answer C is incorrect. A/C system pressures vary dramatically with ambient humidity changes.

   Answer D is correct. A/C system pressures do not vary with cab humidity. However, outside humidity will change A/C system pressures dramatically.

9. A compressor cycling on and off too fast is most commonly caused by which of the following conditions?

   A. Defective compressor clutch

   B. Defective control switch

   C. Overcharged system

   D. Low refrigerant charge

TASK B.1.3

   Answer A is incorrect. A defective clutch will not cycle at all.

   Answer B is incorrect. A defective control switch will cause the system to be inoperative and have no compressor operation.

   Answer C is incorrect. An overcharged system will cause the clutch to remain engaged and the compressor operation will be louder than normal due to the excess pressures.

   Answer D is correct. Low refrigerant charge will cause the clutch to cycle more frequently than normal.

10. What will happen if R-12 comes into contact with a flame?

   A. It will explode.

   B. It will form a non-toxic gas.

   C. It will form chlorine crystals.

   D. It will form phosgene gas.

TASK B.1.5

   Answer A is incorrect. R-12 is not flammable.

   Answer B is incorrect. The gas that is formed is toxic, attacking the nervous system.

   Answer C is incorrect. R-12 does not form chlorine gas crystals in the presence of a flame.

   Answer D is correct. When R-12 is burned it creates phosgene gas. Phosgene gas is dangerous to humans.

11. Technician A says that moisture in an A/C system can combine with the refrigerant to form harmful acids. Technician B says that moisture may be removed from an A/C system by evacuating. Who is correct?

    A. A only

    B. B only

    C. Both A and B

    D. Neither A nor B

    Answer A is incorrect. Technician B is also correct.

    Answer B is incorrect. Technician A is also correct.

    Answer C is correct. Both Technicians are correct. If moisture enters the A/C system, it mixes with the refrigerant to form harmful acids. The method of removing any moisture from the A/C system is to pull a vacuum on the system for at least 30 minutes. Pulling a vacuum on the system, or evacuating it, will cause any moisture to vaporize and be pulled out of the system.

    Answer D is incorrect. Both Technicians are correct.

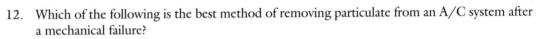

12. Which of the following is the best method of removing particulate from an A/C system after a mechanical failure?

    A. Putting A/C flush solvent in the reverse flow

    B. R-11 flushing

    C. Using an in-line filter

    D. Nitrogen flushing

    Answer A is correct. A/C flush solvent, such as Dura 141, can remove debris from lines and hose assemblies that do not have mufflers in them. If the line or hose assembly has a muffler, then it cannot be flushed and must be replaced. Condensers that are used on R-134a systems have very small passages. In the event of a mechanical failure, the passages of the condenser get clogged and cannot be effectively flushed out. The condenser needs to be replaced.

    Answer B is incorrect. CFC refrigerants must never be used to flush the A/C system.

    Answer C is incorrect. In order for a filter to work, the particulate must circulate through the system, creating the potential for component damage or clogging.

    Answer D is incorrect. Nitrogen flushing will not clean residue from the lines and hoses. However, using pressurized nitrogen is a good method to push the flush solvent through the system.

13. The LEAST LIKELY cause of the high-pressure relief valve tripping open is:

    A. Improper radiator shutter operation.

    B. A clogged condenser.

    C. Inoperative cooling fan clutch.

    D. A defective A/C compressor.

    Answer A is incorrect. A problem that causes the shutters to remain closed with the A/C on could cause very high system pressures and trip the high-pressure relief valve.

    Answer B is incorrect. A clogged condenser could cause the relief valve to trip by causing excessive high-side pressure.

    Answer C is incorrect. An inoperative fan clutch could cause the relief valve to trip by not moving enough air across the condenser.

    Answer D is correct. A defective A/C compressor will not cause the relief valve to trip. If the compressor is defective, it will not be able to develop the high pressure needed to trip the pressure relief valve. The relief valve typically opens up at approximately 450 to 550 psi.

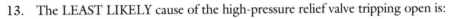

14. A binary pressure switch provides which of the following?

    A. Only low-pressure protection for the A/C system

    B. Only high-pressure protection for the A/C system

    C. Both low- and high-pressure protection for the A/C system

    D. Neither low-pressure nor high-pressure protection for the A/C system

TASK B.2.1

Answer A is incorrect. The binary switch also provides high-pressure protection.

Answer B is incorrect. The binary switch also provides low-pressure protection.

Answer C is correct. The binary switch disables the A/C compressor when the system pressure is too high or too low. The binary switch is usually mounted on the receiver/drier and is used on a TXV system.

Answer D is incorrect. The binary switch does provide protection for the compressor in low- or high-pressure situations.

15. All of the following can cause A/C compressor drive belt misalignment EXCEPT:

    A. A bent compressor mounting bracket holder.

    B. A faulty compressor clutch bearing assembly.

    C. An improperly set compressor clutch air gap.

    D. A faulty idler pulley.

TASK B.2.3

Answer A is incorrect. A bent mounting bracket can cause belt misalignment.

Answer B is incorrect. A faulty clutch bearing can cause belt misalignment.

Answer C is correct. Shimming the clutch plate sets the air gap; the air gap will not affect belt alignment.

Answer D is incorrect. A faulty idler pulley will cause belt misalignment.

16. Technician A says that the best way to ensure that an A/C compressor has the proper amount of lubricant is to drain the compressor and add lubricant to the manufacturer's specifications. Technician B says if you have a doubt about the lubricant level, add an oil charge to the system. Who is correct?

TASK B.2.5

    A. A only

    B. B only

    C. Both A and B

    D. Neither A nor B

Answer A is correct. Only Technician A is correct. This is the proper method of checking the oil level in the compressor.

Answer B is incorrect. Simply adding an oil charge could result in too much lubricant in the system. This could result in elevated system pressures and lower A/C performance.

Answer C is incorrect. Only Technician A is right.

Answer D is incorrect. Technician A is correct.

TASK B.2.6

17. Which of the following items is LEAST LIKELY to be checked and inspected when replacing an A/C compressor?

   A. The piston ring end clearance
   B. The clutch air gap
   C. The number of pulley grooves
   D. The location of the mounting holes

   Answer A is correct. Checking the end clearance would require the compressor to be completely disassembled.

   Answer B is incorrect. The clutch air gap needs to be checked with a feeler gauge during a compressor replacement procedure.

   Answer C is incorrect. The pulley groove count should be the same on the replacement compressor as the old compressor.

   Answer D is incorrect. The compressor mounting locations and brackets should match the old compressor.

TASK B.3.3

18. The A/C condenser has several bent fins and a moderate accumulation of dead insects. Technician A says that the condenser should be replaced or it will cause the high-pressure relief valve to trip. Technician B says that the condenser fins should be straightened and the dead insects removed to optimize A/C system performance. Who is correct?

   A. A only
   B. B only
   C. Both A and B
   D. Neither A nor B

   Answer A is incorrect. The condenser should not be replaced: The fins should be straightened and then the condenser should be cleaned.

   Answer B is correct. Only Technician B is correct. Several bent fins and a moderate accumulation of dead insects will restrict airflow through the condenser enough to cause poor cooling and higher than normal system pressures. It is a good practice to check for condenser restrictions while performing a good preventative maintenance inspection.

   Answer C is incorrect. Only Technician B is correct.

   Answer D is incorrect. Technician B is correct.

TASK B.3.8

19. When attempting to verify a leaking evaporator core, the technician is LEAST LIKELY to sense refrigerant with the detector probe:

   A. At the evaporator case drain.
   B. Over the block type expansion valve.
   C. At the panel vents.
   D. At the defroster vents.

   Answer A is incorrect. The evaporator drain is the most likely place to detect refrigerant leaking from the evaporator core.

   Answer B is correct. The block-type expansion valve is outside of the evaporator case.

   Answer C is incorrect. Refrigerant leaking from the evaporator core could possibly be detected at the panel vents.

   Answer D is incorrect. Refrigerant leaking from the evaporator core could possibly be detected at the defrost vents.

20. Which of the following examples would be the best method of removing a mildew odor from the HVAC duct?

    A. Remove the evaporator case, clean it with a vinegar and water solution, and dry it thoroughly before reinstallation.

    B. Spray a disinfectant into the panel outlets.

    C. Pour a small amount of alcohol into the air intake plenum.

    D. Place an automotive deodorizer under the dash.

TASK B.3.9

    Answer A is correct. Vinegar will kill the mold and mildew that is the source of the odor. Another method to correct a mildew odor is to use a fungicide spray on the evaporator core. This can be applied by removing the blower resistor and spraying the surface of the evaporator core. If this does not correct the problem, the evaporator core should be replaced.

    Answer B is incorrect. This method will simply mask the odor and not eliminate the problem.

    Answer C is incorrect. This method poses a possible fire hazard.

    Answer D is incorrect. This method will simply mask the odor and not eliminate the problem.

21. Technician A says the R-12 A/C service valve has a removable core. Technician B says the R-134a service valve must be rear seated during A/C compressor replacement. Who is correct?

    A. A only

    B. B only

    C. Both A and B

    D. Neither A nor B

TASKS B.2.6, B.3.2

    Answer A is correct. Only Technician A is correct. R-12 A/C service valves have a replaceable Schrader valve.

    Answer B is incorrect. R-134a service valves cannot be rear seated.

    Answer C is incorrect. Only Technician A is correct.

    Answer D is incorrect. Technician A is correct.

22. The A/C high-pressure relief valve shows evidence of slight oil leakage. Technician A says it is necessary to replace the valve and repair the leak. Technician B says that if the valve is just leaking a little oil and not refrigerant, repair is not necessary. Who is correct?

    A. A only

    B. B only

    C. Both A and B

    D. Neither A nor B

TASKS B.1.4, B.3.2

    Answer A is correct. Only Technician A is correct. Any leak from the system must be repaired as quickly as possible.

    Answer B is incorrect. If oil is leaking from the system, refrigerant must also be leaking from the system.

    Answer C is incorrect. Only Technician A is correct.

    Answer D is incorrect. Technician A is correct.

TASK C.1

23. Technician A says that a clogged heater core could cause insufficient heater output. Technician B says that an improperly adjusted coolant control valve cable could cause insufficient heater output. Who is correct?

A. A only

B. B only

C. Both A and B

D. Neither A nor B

Answer A is incorrect. Technician B is also correct.

Answer B is incorrect. Technician A is also correct.

Answer C is correct. Both Technicians are correct. Insufficient heater output could be caused by a clogged heater core or a misadjusted heater control valve. Either problem could cause a low amount of heated water to pass through the core.

Answer D is incorrect. Both Technicians are correct.

TASK C.3

24. Which of the following tools can be used to measure the freeze protection of engine coolant?

A. Litmus strip

B. Manometer

C. Refractometer

D. Opacity meter

Answer A is incorrect. A litmus strip is used to measure relative acidity/alkalinity.

Answer B is incorrect. A manometer is used to measure inlet restriction.

Answer C is correct. Antifreeze protection should be measured with a refractometer. A hydrometer can also be used to check the antifreeze protection.

Answer D is incorrect. An opacity meter is used to measure smoke density.

TASK C.5

25. Which of the following is a typical heavy-duty truck cooling system operating pressure?

A. 3 psi

B. 15 psi

C. 20 psi

D. 30 psi

Answer A is incorrect. A cooling system operating pressure of 3 psi is far too low.

Answer B is correct. Heavy-duty truck cooling systems are typically pressurized to about 10 psi to 15 psi. For every 1 psi of pressure that is added to the cooling system, the boiling point is raised 3° F.

Answer C is incorrect. A cooling system operating pressure of 20 psi is too high and could possibly cause leaks in cooling system components.

Answer D is incorrect. A cooling system operating pressure of 30 psi is very high and would likely cause leaks in cooling system components.

26. Each of the following is an indication that the thermostat has opened EXCEPT:

    A. Coolant flow is visible in the upper radiator tank.

    B. The upper radiator hose is hot.

    C. The lower radiator hose is hot.

    D. The temperature gauge has stabilized in the normal range.

TASK C.7

Answer A is incorrect. Visible coolant flow in the upper radiator tank is an indication that the thermostat has opened.

Answer B is incorrect. The upper radiator hose gets hot when the thermostat opens and hot coolant begins flowing to the radiator.

Answer C is correct. The temperature of the lower radiator hose is not a good indicator of an open thermostat.

Answer D is incorrect. When the thermostat opens, the temperature should stop rising and stabilize in the normal range.

27. Flushing the cooling system does not:

    A. Remove rust from the system.

    B. Remove contaminants from the system.

    C. Increase the life of the cooling system.

    D. Remove acids of combustion from the cooling system.

TASK C.8

Answer A is incorrect. Flushing the cooling system removes rust from the system.

Answer B is incorrect. Flushing the cooling system removes contaminants from the system.

Answer C is incorrect. Flushing the cooling system can increase the life of the system components.

Answer D is correct. Acids of combustion are typically found in the engine oil, not in the coolant.

28. The inside of the windshield has a sticky film. Technician A says to check the engine coolant level. Technician B says the heater core may be leaking. Who is correct?

    A. A only

    B. B only

    C. Both A and B

    D. Neither A nor B

TASK C.2

Answer A is incorrect. Technician B is also correct.

Answer B is incorrect. Technician A is also correct.

Answer C is correct. Both Technicians are correct. A sticky film on the windshield is an indication of an engine coolant leak at the heater core.

Answer D is incorrect. Both Technicians are correct.

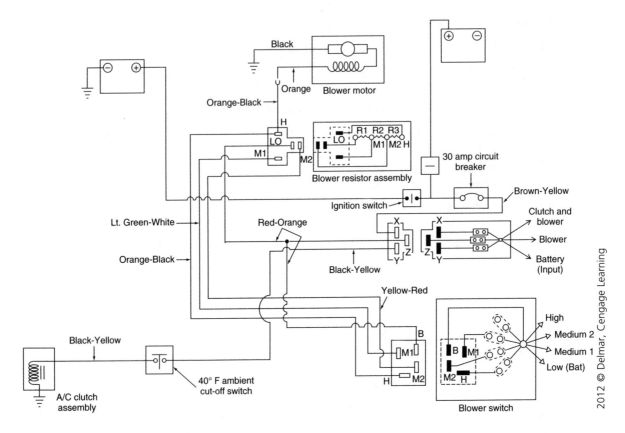

29. Referring to the figure above, all of the following could prevent the A/C clutch from engaging EXCEPT:

TASK D.1

A. A faulty 30 amp circuit breaker.

B. Ambient temperature below 40° F (4.4° C).

C. A stuck closed ambient temperature cut-off switch.

D. An open circuit in the black-yellow wire.

Answer A is incorrect. A faulty circuit breaker will deny power to the system.

Answer B is incorrect. Ambient air temperature below 40° F (4.4° C) will open the ambient cut-off switch, disabling the compressor clutch.

Answer C is correct. If the ambient cut-off switch were stuck closed, the compressor clutch would not be disabled.

Answer D is incorrect. An open black-yellow wire will disable the compressor clutch.

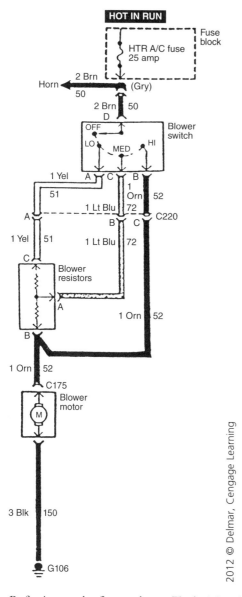

2012 © Delmar, Cengage Learning

30. Referring to the figure above, Technician A says that the light blue wire receives power when the blower switch is turned to the low position. Technician B says that the orange wire receives power when the blower switch is turned to the high position. Who is correct?

TASK D.2

A. A only

B. B only

C. Both A and B

D. Neither A nor B

Answer A is incorrect. The light blue wire does not receive power when the blower is turned to low. The light blue wire does receive power when the blower switch is turned to the medium position.

Answer B is correct. Only Technician B is correct. The orange wire does receive power when the blower switch is turned to the high position. The blower resistor is by-passed when the blower switch is in the high position.

Answer C is incorrect. Only Technician B is correct.

Answer D is incorrect. Technician B is correct.

TASK D.4

31. Technician A says that many trucks are programmed with an engine protection strategy to warn the driver if coolant level and temperature are not within preset parameters. Technician B says that imminent engine shut down can be overridden for a brief period after the alarm to allow the driver to move the vehicle to a safe place. Who is correct?

A. A only

B. B only

C. Both A and B

D. Neither A nor B

Answer A is incorrect. Technician B is also correct.

Answer B is incorrect. Technician A is also correct.

Answer C is correct. Both Technicians are correct. Engine protection systems will first derate the engine and then kill the engine if critical situations develop in the engine, such as low oil pressure or high coolant temperature. Most protection systems will allow the truck to be restarted and moved in case the truck is in a dangerous location, such as on a train track.

Answer D is incorrect. Both Technicians are correct.

TASK D.8

32. Technician A says that an electronic blend door motor uses a pulse-width modulated (PWM) signal to control the position of the door. Technician B says that an electronic mode door motor uses a feedback device to indicate the position of the door. Who is correct?

A. A only

B. B only

C. Both A and B

D. Neither A nor B

Answer A is incorrect. Electronic blend door motors are not PWM-controlled.

Answer B is correct. Only Technician B is correct. Electronic blend door motors use some sort of feedback device. A potentiometer is typically used to signal the HVAC control device about the correct door position.

Answer C is incorrect. Only Technician B is correct.

Answer D is incorrect. Technician B is correct.

TASK D.7

33. Before replacing an HVAC electronic control panel, the technician should:

A. Remove the control cables from the vehicle.

B. Disconnect the batteries.

C. Disassemble the dash panel.

D. Apply dielectric grease to the switch contacts.

Answer A is incorrect. There is no need to remove control cables from the vehicle before replacing any electronic control panel or device that could have solid-state circuitry.

Answer B is correct. The technician should disconnect the batteries before replacing an electronic control panel. This will help prevent a voltage spike when handling the unit. The technician should also ground himself before touching the electronic panel to prevent static electricity from damaging it.

Answer C is incorrect. Some HVAC electronic control panels can be replaced without disassembling the dash panels.

Answer D is incorrect. Nothing should be applied to the switch contacts.

34. What is the LEAST LIKELY cause of the HVAC mode switch functioning incorrectly?

    A. Vacuum leak

    B. Air leak

    C. Broken cable

    D. Open blower resistor

TASK D.7

    Answer A is incorrect. A vacuum leak could cause the HVAC mode switch to not function correctly if the modes are changed using vacuum actuators.

    Answer B is incorrect. An air leak could cause the HVAC mode switch to not function correctly if the modes are changed using air actuators.

    Answer C is incorrect. A broken cable could cause the HVAC mode switch to not function correctly if the modes are changed by cable connections.

    Answer D is correct. An open blower resistor will not affect mode switch operation. A resistor problem usually affects the blower motor in the low speeds.

35. A whistling noise coming from under the dash while the HVAC system is being operated with the blower on HI could indicate which of the following?

    A. A misaligned duct

    B. A defective vacuum actuator

    C. An improperly adjusted mode door cable

    D. A poor electrical connection to the blend door motor

TASK D.8

    Answer A is correct. A misaligned duct could cause a whistling noise. Other possible causes of a whistling noise might be a cracked case or foreign items, such as paper or leaves, in the ducts or blower cage.

    Answer B is incorrect. A defective actuator will not cause this symptom. If an actuator were bad, the air would not come out at the correct place.

    Answer C is incorrect. An improperly adjusted cable will not cause this symptom. If the mode door cable were misadjusted, the air would not come out at the correct place.

    Answer D is incorrect. A poor connection will not cause this symptom. A poor connection at the blend door motor would result in no control of the temperature of the air.

36. A truck is being diagnosed for an ATC system problem. The blend air door constantly moves back and forth. Which of the following conditions would be the most likely cause of this problem?

    A. Binding blend air door

    B. Defective actuator motor

    C. Improperly adjusted ATC sensor cable

    D. Defective feedback device

TASK D.9

    Answer A is incorrect. A binding door will not cause the actuator to hunt for the proper position. This problem would cause a popping noise in the dash and poor operation of the binding door.

    Answer B is incorrect. A faulty motor will not cause the actuator to hunt for the desired position. The system would have no change in temperature if the blend door actuator was bad.

    Answer C is incorrect. Most ATC actuators do not use a cable to connect to the door. Typically, the actuator directly attaches to the door and the feedback sensor is located inside the actuator.

    Answer D is correct. A faulty feedback device can cause the blend air door to hunt for the desired position. The feedback device is typically a potentiometer that uses three wires. One wire is the input voltage from the controller; a second wire is the return ground from the controller; the third wire is the signal wire to the controller.

TASK B.1.8

37. To avoid mixing refrigerants, Technician A says that the R-134a service hose fittings are different from R-12 fittings. Technician B says that R-12 containers may be either white or yellow, but R-134a containers are sky blue. Who is correct?

  A. A only

  B. B only

  C. Both A and B

  D. Neither A nor B

Answer A is incorrect. Technician B is also correct.

Answer B is incorrect. Technician A is also correct.

Answer C is correct. Both Technicians are correct. The containers are color coded and the fittings are different to avoid mixing refrigerants.

Answer D is incorrect. Both Technicians are correct.

TASK B.1.12

38. The refrigerant is being recovered from a late-model truck with an A/C recovery/recycling machine. Technician A says that some recycling machines filter the refrigerant as the recovery process is taking place. Technician B says that some recycling machines filter the refrigerant as the recharge process is taking place. Who is correct?

  A. A only

  B. B only

  C. Both A and B

  D. Neither A nor B

Answer A is incorrect. Technician B is also correct.

Answer B is incorrect. Technician A is also correct.

Answer C is correct. Both Technicians are correct. Some A/C recovery/recycling machines filter the refrigerant during recovery and some A/C machines filter the refrigerant during the recharge process. In addition, one equipment manufacturer filters the refrigerant during both processes.

Answer D is incorrect. Both Technicians are correct.

TASK B.1.6

39. Before a portable container is used to transfer recycled R-12, it must be evacuated to at least:

  A. 20 in. Hg (67.7 kPa)

  B. 22 in. Hg (74.5 kPa)

  C. 27 in. Hg (91.4 kPa)

  D. 12 in. Hg (40.6 kPa)

Answer A is incorrect. A vacuum of 20 in. Hg (67.7 kPa) is not sufficient to ensure that all moisture has been removed from the tank.

Answer B is incorrect. A vacuum of 22 in. Hg (74.5 kPa) is not sufficient to ensure that all moisture has been removed from the tank.

Answer C is correct. A vacuum of 27 in. Hg (91.4 kPa) is required to remove all of the moisture from the container.

Answer D is incorrect. A vacuum of 12 in. Hg (40.6 kPa) is not sufficient to ensure that all moisture has been removed from the tank.

40. What is the source of most non-condensable gases in refrigerant?

    A. Acid
    B. Air
    C. Moisture
    D. Oil

TASK B.1.11

Answer A is incorrect. Any acids that form in the A/C system are created due to moisture and corrosion and are usually in a liquid state.

Answer B is correct. Air is not condensable at A/C system pressures. To identify if a vessel of refrigerant has excessive air, a technician can check the pressure in the vessel and compare this pressure with the pressure/temperature chart.

Answer C is incorrect. Moisture in the A/C system only evaporates when the system is drawn under a vacuum and is, therefore, normally in a liquid state.

Answer D is incorrect. Refrigeration oil is normally in a liquid state.

PREPARATION EXAM 2—ANSWER KEY

1.	B	21.	B
2.	A	22.	A
3.	D	23.	B
4.	D	24.	A
5.	D	25.	A
6.	D	26.	A
7.	A	27.	D
8.	B	28.	D
9.	B	29.	C
10.	D	30.	C
11.	B	31.	C
12.	C	32.	A
13.	C	33.	A
14.	B	34.	D
15.	D	35.	B
16.	A	36.	C
17.	B	37.	B
18.	A	38.	C
19.	A	39.	B
20.	B	40.	D

PREPARATION EXAM 2—EXPLANATIONS

TASK A.1

1. A truck's A/C system does not cool when driving at slow speeds. Technician A says that the truck needs a fan clutch. Technician B says that the truck needs to be road tested to see if the A/C cools well at highway speeds. Who is correct?

   A. A only

   B. B only

   C. Both A and B

   D. Neither A nor B

   Answer A is incorrect. A technician should never just use the repair order to diagnose a vehicle. Each concern on the repair order should be verified to make sure that the driver is accurate in the concern.

   Answer B is correct. Only Technician B is correct. A technician should never assume that any vehicle has the problem that is on the repair order. It is always wise to verify the complaint. This vehicle should be road tested to gather as much information as possible before beginning the diagnosis.

   Answer C is incorrect. Only Technician B is correct.

   Answer D is incorrect. Technician B is correct.

2. Technician A says that a refrigerant identifier can be used to test for air. Technician B says that a flow tester can be used to test for air. Who is correct?

TASK B.1.11

   A. A only
   B. B only
   C. Both A and B
   D. Neither A nor B

   Answer A is correct. Only Technician A is correct. Most refrigerant identifiers will detect the percentage of air that is mixed with the refrigerant.

   Answer B is incorrect. A flow tester is used to detect the presence of sealer that could be mixed with the refrigerant.

   Answer C is incorrect. Only Technician A is correct.

   Answer D is incorrect. Technician A is correct.

3. A hissing sound is heard under the hood area of a truck after the engine has been turned off. The noise stops after about one minute and only happens when the air conditioner or defroster has been used. Technician A says that the engine cooling system is equalizing. Technician B says that the alternator can produce this noise. Who is correct?

TASK A.2

   A. A only
   B. B only
   C. Both A and B
   D. Neither A nor B

   Answer A is incorrect. The cooling system does not make a noise like this for one minute after the engine is shut off.

   Answer B is incorrect. The alternator should not make any noises after the engine is shut off.

   Answer C is incorrect. Neither Technician is correct.

   Answer D is correct. Neither Technician is correct. The noise is likely the A/C system equalizing and is a normal occurrence on trucks.

4. A truck technician discovers a heater core leak in the sleeper that requires replacement. Technician A says that it is important to add the proper amount of refrigeration oil before installation. Technician B says that a PAG-based lubricant is used in modern heater cores. Who is correct?

TASK C.11

   A. A only
   B. B only
   C. Both A and B
   D. Neither A nor B

   Answer A is incorrect. Oil is not added to the heater core prior to installation.

   Answer B is incorrect. Oil is not added to the heater core prior to installation.

   Answer C is incorrect. Neither Technician is correct.

   Answer D is correct. Neither Technician is correct. Both Technicians are stating that refrigerant oil needs to be added during heater core replacement; this is not required.

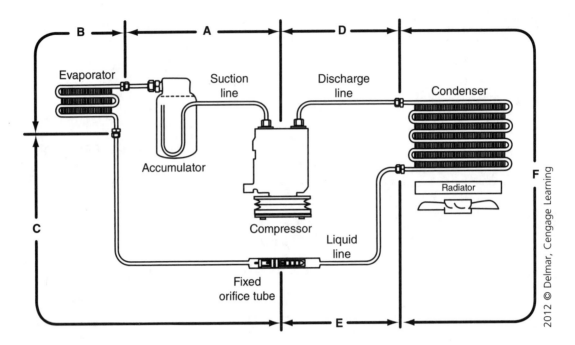

TASK A.3

5. Referring to the figure, Section D is found to be very warm during a performance test of the A/C system on a 90° F (32.2° C) day. Technician A says that a faulty compressor clutch coil could cause this problem. Technician B says that a restricted orifice tube could cause this problem. Who is correct?

    A. A only

    B. B only

    C. Both A and B

    D. Neither A nor B

Answer A is incorrect. A faulty compressor clutch coil would not cause the discharge line to be hot. The line would be at ambient temperature if the clutch coil were faulty.

Answer B is incorrect. A restricted orifice tube would not cause the discharge line to be hot. A restricted orifice tube would cause the compressor to short cycle and the A/C performance would be reduced.

Answer C is incorrect. Neither Technician is correct.

Answer D is correct. Neither Technician is correct. Section D in the figure is the discharge line that contains high-pressure vaporized refrigerant. This part of the refrigeration system will always be the hottest part of the system. The condition described in the question is a normal condition.

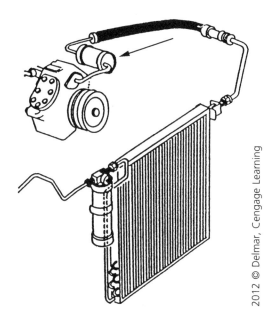

2012 © Delmar, Cengage Learning

6. Referring to the figure, if the component indicated by the arrow were not used, what would be the A/C performance result?

A. Higher A/C high-side pressure

B. Improperly filtered refrigerant

C. Evaporator icing

D. Noisier A/C operation

TASKS A.3, A.4

Answer A is incorrect. The component shown is an A/C muffler; it will have no effect on A/C circuit pressures.

Answer B is incorrect. The component shown is an A/C muffler; it will have no effect on refrigerant filtration.

Answer C is incorrect. The component shown is an A/C muffler; it will not affect evaporator icing.

Answer D is correct. The component shown is an A/C muffler; if it were not used, A/C operation would be noisier.

7. Poor cooling from an A/C system that uses a TXV valve can be caused by all of the following EXCEPT:

A. A fan clutch that is always engaged.

B. An improperly adjusted blend door cable.

C. A low refrigerant charge.

D. A refrigerant overcharge.

TASK B.1.1

Answer A is correct. If the fan clutch is always engaged, it will enhance the operation of the A/C system.

Answer B is incorrect. An improperly adjusted blend door cable will cause poor A/C performance.

Answer C is incorrect. A low refrigerant charge will cause poor A/C system performance.

Answer D is incorrect. A refrigerant overcharge will cause poor A/C system performance.

TASK C.2

8. With the HVAC system in DEF mode, the blower on HI, and the temperature control on COLD, the bottom of the windshield fogs up on the outside because:

   A. The evaporator case drain is clogged.
   B. The heater core leaks.
   C. The evaporator core is iced up.
   D. The cold windshield causes moisture to condense from the outside air.

   Answer A is incorrect. A clogged evaporator drain will cause fogging on the inside of the windshield, accompanied by a mildew smell.

   Answer B is incorrect. A leaking heater core will cause fogging on the inside of the windshield, accompanied by a sweet smell.

   Answer C is incorrect. An iced-up evaporator core will not cause the windshield to fog. This problem would result in poor A/C cooling.

   Answer D is correct. Under these conditions, a cold windshield causes moisture to condense from the outside air.

TASK B.1.4

9. Where should the probe of an electronic lead detector be placed in order to most effectively find a refrigerant leak?

   A. Directly above the suspected leak area
   B. Directly below the suspected leak area
   C. Six inches upstream from the suspected leak area
   D. Six inches downstream from the suspected leak area

   Answer A is incorrect. Refrigerants are heavier than air and will not be effectively detected above the suspected leak.

   Answer B is correct. Refrigerants are heavier than air and will be most effectively detected below a suspected leak.

   Answer C is incorrect. Six inches upstream from the leak would cause small leaks to go undetected.

   Answer D is incorrect. Six inches downstream from the leak would cause small leaks to go undetected.

TASK B.1.7

10. Technician A says that A/C refrigerant hoses can be flushed with R-134a. Technician B says that the condenser can be flushed with R-12. Who is correct?

    A. A only
    B. B only
    C. Both A and B
    D. Neither A nor B

    Answer A is incorrect. R-134a should not be used to flush refrigerant hoses.

    Answer B is incorrect. R-12 should not be used to flush the condenser.

    Answer C is incorrect. Neither Technician is correct.

    Answer D is correct. Neither Technician is correct. Neither R-12 nor R1-34a should be used to openly flush components of the A/C system.

11. Technician A says you can complete the high-side charging procedure with the engine running. Technician B says if liquid refrigerant enters the compressor, damage will result to the compressor. Who is correct?

TASK B.1.8

    A. A only
    B. B only
    C. Both A and B
    D. Neither A nor B

    Answer A is incorrect. If the technician charges into the high side with the engine running, high-pressure refrigerant may be forced into the container or charging machine.

    Answer B is correct. Only Technician B is correct. If liquid refrigerant enters the compressor, then the compressor will be damaged because liquid cannot be compressed.

    Answer C is incorrect. Only Technician B is correct.

    Answer D is incorrect. Technician B is correct.

12. An electric condenser fan may be controlled by any of the following EXCEPT:

TASK D.5

    A. The engine control unit.
    B. The chassis management unit.
    C. A manual switch.
    D. An electronic relay.

    Answer A is incorrect. The engine control unit can control an electric condenser fan motor.

    Answer B is incorrect. The body control unit can sometimes control an electric condenser fan motor.

    Answer C is correct. A manual switch is typically not used to control an electric condenser fan.

    Answer D is incorrect. An electric condenser fan motor can be controlled by an electronic relay.

13. The low-pressure cut-out switch senses pressure in the:

TASK B.2.2

    A. System high side.
    B. Atmosphere.
    C. System low side.
    D. Cab/sleeper.

    Answer A is incorrect. The low-pressure cut-out switch is located in the low side of the system.

    Answer B is incorrect. An aneroid device senses inlet manifold pressure, which will be greater than atmospheric pressure in a turbocharged engine. Atmospheric pressure could be sensed by a barometric pressure sensor.

    Answer C is correct. The low-pressure cut-out switch (also called the pressure cycling switch) is mounted in the low side of the A/C system and opens when the pressure drops below 20 to 25 psi.

    Answer D is incorrect. The low-pressure switch is typically not located in the cab or sleeper.

14. Which of the following statements about coolant control valve replacement in an ATC system is true?

TASK C.10

    A. Refrigerant must be recovered before the coolant control valve is replaced.
    B. Heater hoses connected to the coolant control valve may be clamped during the replacement procedure to maintain coolant in the system.
    C. Access to the coolant control valve is gained through the fresh air door.
    D. The coolant control valve is an integral part of the heater hose.

    Answer A is incorrect. The refrigerant does not need to be recovered from the system before the valve is replaced.

    Answer B is correct. The heater hoses connected to the coolant control valve may be clamped during the replacement procedure to maintain coolant in the system. Many tool companies sell a tool to perform this function called hose pinching pliers.

    Answer C is incorrect. The coolant control valve is located in front of the firewall.

    Answer D is incorrect. The coolant control valve is an individually replaceable component.

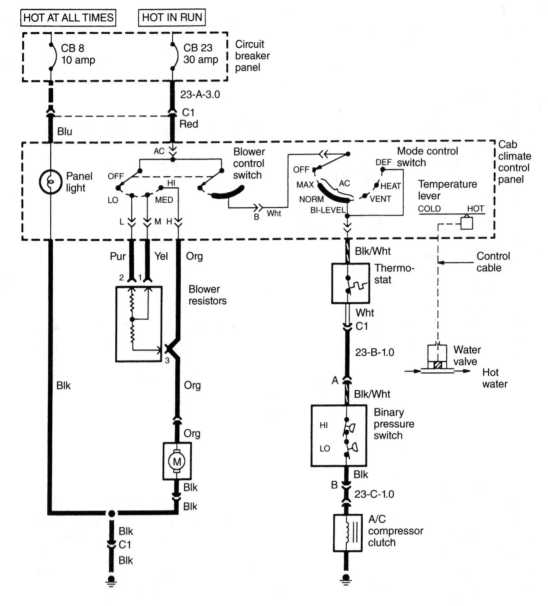

TASK B.2.4

15. Referring to the figure, all of the following could prevent A/C compressor clutch engagement EXCEPT:

A. A defective blower switch.

B. An open binary pressure switch.

C. A poor connection at the A/C thermostat.

D. An open circuit breaker at CB 8.

Answer A is incorrect. A defective blower switch could prevent clutch engagement by not allowing current to flow from the white wire to the mode control switch.

Answer B is incorrect. An open binary pressure control switch will prevent clutch engagement because it is in the path to the A/C clutch.

Answer C is incorrect. A poor connection at the thermostat switch could prevent clutch engagement.

Answer D is correct. If this circuit breaker opens, it affects only the panel light and will not prevent clutch engagement.

16. The compressor discharge valve is designed to:

    A. Open after vaporized refrigerant is compressed, allowing the refrigerant to move to the condenser.

    B. Open before vaporized refrigerant is compressed, allowing the refrigerant to move to the evaporator.

    C. Regulate system variable pressure.

    D. Regulate A/C system temperature.

TASK B.2.6

    Answer A is correct. The compressor discharge valve opens after the pressure compresses vaporized refrigerant, allowing the refrigerant to move to the condenser to change from a high-pressure gas into a high-pressure liquid.

    Answer B is incorrect. If the discharge valve were to open before vaporized refrigerant was compressed, it would not properly raise the refrigerant pressure to allow a change from gas to a liquid while in the condenser. In addition, refrigerant does not go from the compressor to the evaporator.

    Answer C is incorrect. The reed valves just open and close; they do not regulate a variable pressure.

    Answer D is incorrect. Reed valves cannot sense A/C system temperature. Reed valves are just mechanical valves that separate the low side and high side of the A/C system.

17. Which of the following conditions would be LEAST LIKELY to cause a knocking sound when the A/C compressor is engaged?

    A. Broken compressor mounting bolt

    B. Restricted suction line

    C. Discharge line rubbing a metal bracket

    D. Loose compressor mounting bolts

TASK B.2.3

    Answer A is incorrect. A broken compressor mounting bolt could cause the compressor to produce a knocking sound when engaged because it would not be secured properly.

    Answer B is correct. A restricted suction line would not likely cause a knocking sound when the compressor is engaged. A restricted suction line could cause the compressor to short cycle due to the blockage of refrigerant flow.

    Answer C is incorrect. A discharge line rubbing a metal bracket could produce a knocking noise when the compressor is engaged due to the vibration of the line being delivered into the metal bracket.

    Answer D is incorrect. Loose compressor mounting bolts could cause the compressor to produce a knocking sound when engaged because the compressor would not be secured properly.

18. A signal to the engine control module (ECM) from which of the following sensors could cause the A/C compressor to disengage?

    A. Engine coolant temperature (ECT) sensor

    B. Intake air temperature (IAT) sensor

    C. Heated oxygen sensor (HO$_2$S)

    D. Cooling fan control sensor

TASK D.4

    Answer A is correct. If the coolant temperature rises to a predetermined level, the ECM will disengage the compressor clutch. This helps an overheating engine by allowing the condenser to cool down and by removing the mechanical load of driving the compressor.

    Answer B is incorrect. The IAT sensor signal is not used to control compressor operation.

    Answer C is incorrect. The oxygen sensor signal is not used to control compressor operation.

    Answer D is incorrect. There is no cooling fan sensor.

TASK B.3.2

19. Technician A says that some systems have an in-line filter. Technician B says that petroleum jelly should be used to lubricate new A/C o-rings. Who is correct?

    A. A only
    B. B only
    C. Both A and B
    D. Neither A nor B

    Answer A is correct. Only Technician A is correct. An in-line filter is sometimes used in the line from the condenser to the evaporator.

    Answer B is incorrect. Petroleum jelly is not a recommended lubricant for A/C o-rings. The recommended lubricant for all A/C o-rings is mineral oil.

    Answer C is incorrect. Only Technician A is correct

    Answer D is incorrect. Technician A is correct.

TASK C.10

20. Technician A says that the best way to test a vacuum-operated coolant control valve is to disconnect the vacuum hose and observe whether coolant flows through it. Technician B says applying and releasing vacuum and checking if the valve arm moves freely both ways is a way of testing a vacuum-operated coolant control valve. Who is correct?

    A. A only
    B. B only
    C. Both A and B
    D. Neither A nor B

    Answer A is incorrect. Some coolant control valves close when vacuum is removed.

    Answer B is correct. Only Technician B is correct. This process allows the technician to check the specific coolant flow valve operation. The valve should hold the vacuum for at least one minute without leaking down.

    Answer C is incorrect. Only Technician B is correct.

    Answer D is incorrect. Technician B is correct.

TASK B.3.7

21. What is the function of the orifice tube?

    A. It allows water to drain from the evaporator case.
    B. It allows refrigerant to be metered into the evaporator.
    C. It regulates refrigerant flow through the condenser.
    D. It regulates airflow through the evaporator.

    Answer A is incorrect. The orifice tube is located in the evaporator core inlet.

    Answer B is correct. The orifice tube allows low-pressure liquid to be metered into the evaporator. Since the pressure is much lower after the orifice tube, the low-pressure refrigerant starts to boil and take on heat from the surrounding surface and air.

    Answer C is incorrect. The orifice tube does not affect the refrigerant flow through the condenser.

    Answer D is incorrect. The orifice tube does not affect airflow through the evaporator.

22. Technician A says that some R-134a service ports have replaceable valve cores. Technician B says that a leaking R-134a service port cap will prevent refrigerant from escaping when the valve core leaks. Who is correct?

TASK B.3.2

    A. A only
    B. B only
    C. Both A and B
    D. Neither A nor B

    Answer A is correct. Only Technician A is correct. The valve core can be replaced on many R-134a service ports. On trucks that do not have serviceable cores, the line can be replaced to repair the leak.

    Answer B is incorrect. The service port cap will not prevent a refrigerant leak if the valve core is leaking. The cap can slow down the leak, but the refrigerant will soon escape and cause the system to not operate correctly.

    Answer C is incorrect. Only Technician A is correct.

    Answer D is incorrect. Technician A is correct.

23. Which of the following is a normal range for the low-side gauge while performing a performance test?

TASK B.1.3

    A. 5–10 psi
    B. 25–45 psi
    C. 60–80 psi
    D. 180–205 psi

    Answer A is incorrect. A low-side pressure range of 5–10 psi is extremely low and is indicative of a clogged orifice tube or low refrigerant charge.

    Answer B is correct. A low-side pressure range of 25–45 psi is normal.

    Answer C is incorrect. A low-side pressure range of 60–80 psi is very high.

    Answer D is incorrect. A range of 180–205 psi is typical for high-side pressure.

24. When a cooling system is pressure tested, there are no obvious external leaks, but the system cannot maintain pressure. Which of the following is the most likely cause of this problem?

TASK C.3

    A. Leaking evaporator
    B. Defective heater valve
    C. Stuck open thermostat
    D. Blown head gasket in the engine

    Answer A is incorrect. The evaporator is not a cooling system component.

    Answer B is incorrect. A defective heater valve would have been found during the external inspection.

    Answer C is incorrect. A stuck open thermostat will cause over-cooling of the engine.

    Answer D is correct. A blown head gasket is the most likely cause of an internal engine coolant leak.

TASK C.4

25. Cooling and heating system hoses should be replaced for any of the following reasons EXCEPT:

   A.   They leak at the hose clamps.
   B.   They are cracked.
   C.   They show signs of bulging.
   D.   They feel spongy.

   Answer A is correct. Cooling and heating system hoses do not need to be replaced because they leak at the hose clamps; usually tightening the clamps repairs the leak.

   Answer B is incorrect. Cracked hoses must be replaced.

   Answer C is incorrect. Bulging hoses must be replaced.

   Answer D is incorrect. Spongy hoses must be replaced.

TASK C.6

26. A growling noise is coming from the water pump. Technician A says that the water pump bearing could be the cause. Technician B says that the impeller has been eroded by cavitation. Who is correct?

   A.   A only
   B.   B only
   C.   Both A and B
   D.   Neither A nor B

   Answer A is correct. Only Technician A is correct. A growling noise from the water pump indicates a worn bearing. If the pump is accessible at all, a technician can use a listening device, such as a stethoscope or a yardstick, and contact the pump housing to pick up the vibration.

   Answer B is incorrect. Cavitation damage will not cause this symptom. Overheating due to the lack of water circulation will result from a damaged water pump impeller.

   Answer C is incorrect. Only Technician A is correct.

   Answer D is incorrect. Technician A is correct.

TASK B.3.5

27. Which of the following is the LEAST LIKELY function of the accumulator/drier that is used on orifice tube systems?

   A.   Removes moisture from the refrigerant and stores it in a desiccant pouch
   B.   Stores refrigerant vapor and oil
   C.   Allows only refrigerant vapor and oil to be sent to the compressor
   D.   Prevents vaporized refrigerant from exiting to the thermal expansion valve

   Answer A is incorrect. The accumulator removes moisture from the refrigerant and stores it in a desiccant pouch.

   Answer B is incorrect. The accumulator is located in the suction line between the evaporator core and the compressor. The accumulator stores refrigerant vapor and oil until the compressor pulls the mixture to be compressed.

   Answer C is incorrect. The accumulator/drier is designed with the exit point at the top of the component, which only allows refrigerant vapor and oil to be sent to the compressor.

   Answer D is correct. The receiver/drier is the device that prevents vaporized refrigerant from exiting to the thermal expansion valve. The receiver/drier is located in the high-pressure liquid line.

28. A sweet odor comes from the panel vents while operating the A/C in NORMAL mode. Technician A says that the smell could be R-12 in the cab, so the truck must have a leaking evaporator. Technician B says the heater core could be leaking. Who is correct?

TASK A.3

    A. A only

    B. B only

    C. Both A and B

    D. Neither A nor B

Answer A is incorrect. R-12 is an odorless gas and would not cause a sweet odor.

Answer B is correct. Only Technician B is correct. The odor of leaking antifreeze can be drawn from the evaporator case even when operating in an A/C mode. If the heater core were causing the odor, the technician would smell a sweet aroma with the HVAC controls set at any setting. Other signs of a leaking heater core might be coolant in the floorboard, a steamed up windshield, or coolant leaking from the HVAC housing drain tube.

Answer C is incorrect. Only Technician B is correct.

Answer D is incorrect. Technician B is correct.

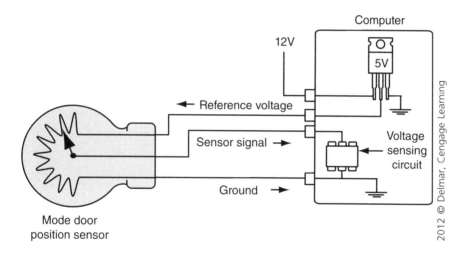

29. Referring to the figure, Technician A says that the computer sends a 5 volt reference to the mode door position switch. Technician B says that the sensor signal varies as the mode door position changes. Who is correct?

TASK D.1

    A. A only

    B. B only

    C. Both A and B

    D. Neither A nor B

Answer A is incorrect. Technician B is also correct.

Answer B is incorrect. Technician A is also correct.

Answer C is correct. Both Technicians are correct. The computer sends a 5 volt reference voltage to the mode door position switch. The signal voltage varies as the mode door position changes. Many electric actuators use potentiometers like this one to provide feedback to the computer concerning door position.

Answer D is incorrect. Both Technicians are correct.

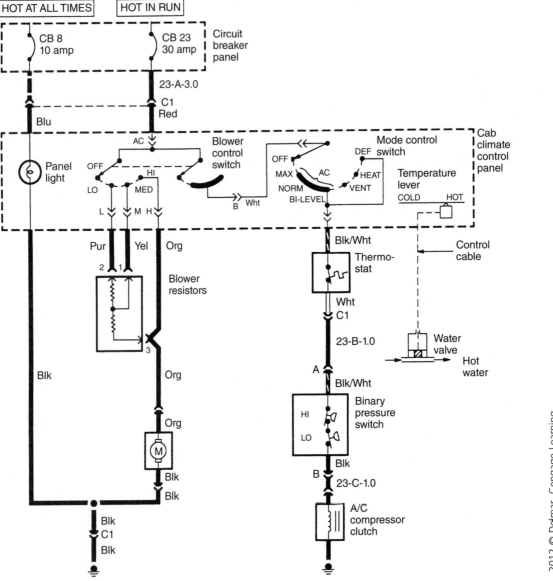

TASK D.2

30. Referring to the figure, which of the following statements is LEAST LIKELY to cause the blower motor to be totally inoperative?

    A. Circuit breaker 23 is open.
    B. The blower motor connector is broken.
    C. The binary pressure switch is open.
    D. The red wire at connector C1 is open.

    Answer A is incorrect. All HVAC functions would be inoperative if circuit breaker C23 were open.

    Answer B is incorrect. The blower motor would be inoperative if the blower motor connector were broken.

    Answer C is correct. An open binary switch would cause the A/C compressor clutch to be inoperative, but it would not cause a problem for the blower motor.

    Answer D is incorrect. An open red wire at connector C1 would cause all HVAC functions to be inoperative.

31. Technician A says that the binary pressure switch prevents compressor operation if the refrigerant charge has been lost or ambient temperature is too cold. Technician B says that the binary pressure switch turns off the compressor if the system pressure is too high. Who is correct?

TASK D.3

    A. A only

    B. B only

    C. Both A and B

    D. Neither A nor B

Answer A is incorrect. Technician B is also correct.

Answer B is incorrect. Technician A is also correct.

Answer C is correct. Both Technicians are correct. The binary pressure switch prevents compressor operation if the refrigerant charge has been lost or ambient temperature is too cold. The binary switch also protects the system from excessive pressure.

Answer D is incorrect. Both Technicians are correct.

32. How much refrigerant oil should be in a typical A/C condenser?

TASK B.3.1

    A. 1 ounce

    B. 5 ounces

    C. 7 ounces

    D. 11 ounces

Answer A is correct. The condenser generally holds 1 ounce of refrigeration oil. Always refer to the manufacturer's recommendations before adding oil to A/C components.

Answer B is incorrect. This amount is too large.

Answer C is incorrect. This amount is too large.

Answer D is incorrect. This amount is too large.

33. Which lubricant is recommended for R-134a mobile A/C systems?

TASK B.1.2

    A. Polyalkylene glycol-based (PAG-based) lubricant

    B. Mineral-based petroleum lubricant

    C. DEXRON or DEXRON II lubricant

    D. Type C-3 SAE 30 based oil lubricant

Answer A is correct. PAG oil is specifically formulated to be compatible with R-134a. PAG oil is a synthetic-based lubricant that is light enough to be circulated by 134a systems.

Answer B is incorrect. R-12 systems use mineral-based lubricant.

Answer C is incorrect. DEXRON is not compatible with R-134a and will damage A/C system components.

Answer D is incorrect. SAE 30 will damage A/C system components.

34. Which of the following is the LEAST LIKELY cause of a binding temperature control cable?

TASK D.8

    A. A kinked cable housing

    B. Corrosion in the cable housing

    C. A deformed or over-tightened cable clamp

    D. A defective mode door

Answer A is incorrect. A kinked cable housing will cause the cable to bind.

Answer B is incorrect. Corrosion in the cable housing will cause the cable to bind.

Answer C is incorrect. A deformed or over-tightened cable clamp will cause the cable to bind.

Answer D is correct. The temperature control cable does not control the mode doors.

TASK D.1

35. Which of the following conditions would most likely result from over-tightening the fasteners on an air actuator?

   A. Ruptured diaphragm
   B. Stripped threads
   C. Deformed linkage
   D. Air leak

Answer A is incorrect. The screws do not contact the diaphragm.

Answer B is correct. Over-tightening the mounting screws can result in stripped threads. Most HVAC cases are made of plastic, so it is important to not use too much force when servicing the components in/on the case.

Answer C is incorrect. Over-tightening the screws will have no effect on the linkage.

Answer D is incorrect. Over-tightened screws are not likely to cause an air leak.

TASK B.2.3

36. A heavy truck is being diagnosed for a belt problem. What is the most likely cause for a serpentine drive belt to slip under heavy loads?

   A. Surface cracks on the inside of the belt
   B. Belt stretched 1/4 inch
   C. Seized tensioner
   D. Glazed belt

Answer A is incorrect. Surface cracks on the inside of a serpentine belt are a sign of age, but they will not cause the belt to slip.

Answer B is incorrect. Stretching the belt 1/4 inch is not enough to cause a serpentine belt to slip. The spring-loaded tensioner will allow for some belt stretch.

Answer C is correct. A seized belt tensioner does not continue to adjust for belt wear and stretch.

Answer D is incorrect. A glazed belt could cause belt noise at various times, but it does not typically cause the belt to slip.

TASK B.1.12

37. Technician A says that anyone who purchases R-134a must maintain records for three years indicating the name and address of the supplier. Technician B says the supplier must maintain sales records of refrigerant purchases. Who is correct?

   A. A only
   B. B only
   C. Both A and B
   D. Neither A nor B

Answer A is incorrect. The purchaser is under no legal obligation to maintain records.

Answer B is correct. Only Technician B is correct. Refrigerant suppliers must maintain records about all facilities to which refrigerant is sent. The seller must retain these sales records for three years.

Answer C is incorrect. Only Technician B is correct.

Answer D is incorrect. Technician B is correct.

38. Which process reduces contaminants in used refrigerant by using oil separation and filter core driers?

    A. Restoration
    B. Recovery
    C. Recycling
    D. Reclamation

TASKS B.1.5, B.1.9

Answer A is incorrect. Restoration is a term generally applied to old buildings.

Answer B is incorrect. Recovery is the process by which refrigerant is removed from a system.

Answer C is correct. The recycling process reduces contaminants used in refrigerant by using oil separation and filter core driers.

Answer D is incorrect. Reclamation is an industrial process by which refrigerant is restored to its original condition by an outside source.

39. A refillable A/C refrigerant cylinder is considered full when it reaches what capacity by weight?

    A. 50 percent
    B. 60 percent
    C. 70 percent
    D. 80 percent

TASK B.1.10

Answer A is incorrect. This level is too low.

Answer B is correct. The tank must not be filled beyond the 60 percent of the gross weight rating. Some room must be left in the cylinder to make room if the temperature is raised.

Answer C is incorrect. This level is too high.

Answer D is incorrect. This level is too high.

40. There is a growling or rumbling noise at the A/C compressor with the compressor clutch engaged or disengaged. Technician A says the compressor bearing is defective. Technician B says the clutch bearing is defective. Who is correct?

    A. A only
    B. B only
    C. Both A and B
    D. Neither A nor B

TASKS A.2, B.2.4

Answer A is incorrect. If the compressor bearing were the cause, the noise would only be present with the clutch engaged.

Answer B is incorrect. A defective clutch bearing causes a growling noise with the clutch disengaged, but goes away when the clutch is turned on. The reason the noise goes away is that both parts of the bearing are being turned at the same speed when the clutch is engaged.

Answer C is incorrect. Neither Technician is correct.

Answer D is correct. Neither Technician is correct. The growling or rumbling noise would have to be coming from something in motion whether the A/C is turned on or off. Items that could possibly cause this problem would include the water pump, an idler bearing, an alternator bearing, or possibly a loose mounting bracket in the drive belt system.

PREPARATION EXAM 3—ANSWER KEY

1.	A	21.	C
2.	C	22.	C
3.	C	23.	D
4.	A	24.	B
5.	D	25.	A
6.	D	26.	B
7.	C	27.	D
8.	A	28.	B
9.	B	29.	D
10.	C	30.	A
11.	D	31.	B
12.	C	32.	B
13.	C	33.	C
14.	C	34.	B
15.	A	35.	A
16.	B	36.	C
17.	A	37.	A
18.	A	38.	B
19.	D	39.	B
20.	B	40.	D

PREPARATION EXAM 3—EXPLANATIONS

TASK A.1

1. A truck is being diagnosed for an A/C problem. Technician A says that the driver's complaint should be closely inspected in order to completely understand the problem. Technician B says that a road test is never necessary when working with A/C problems. Who is correct?

    A. A only
    B. B only
    C. Both A and B
    D. Neither A nor B

    Answer A is correct. Only Technician A is correct. It is very important to closely read the driver's complaint in order to completely understand the problem.

    Answer B is incorrect. Road tests are often necessary when troubleshooting A/C problems on trucks.

    Answer C is incorrect. Only Technician A is correct.

    Answer D is incorrect. Technician A is correct.

2. During an HVAC performance test, the technician hears the A/C compressor clutch slip briefly upon engagement. The most likely cause is:

    A. A worn out compressor clutch coil.

    B. A defective A/C compressor clutch relay.

    C. The compressor clutch air gap is too large.

    D. The compressor clutch bearing is worn.

TASKS A.2, A.4

    Answer A is incorrect. If the compressor clutch coil is defective, then the clutch will not engage at all.

    Answer B is incorrect. A defective relay will prevent the clutch from engaging.

    Answer C is correct. If the air gap is too large, the clutch may slip briefly upon engagement.

    Answer D is incorrect. A worn clutch bearing may cause a loud clutch engagement, but a slipping clutch would not.

3. Technician A says that a heater core leak could cause a coolant smell to be present in the cab area. Technician B says that a leaking heater core could cause coolant to leak onto the floorboard area of the truck. Who is correct?

    A. A only

    B. B only

    C. Both A and B

    D. Neither A nor B

TASKS A.3, C.3

    Answer A is incorrect. Technician B is also correct.

    Answer B is incorrect. Technician A is also correct.

    Answer C is correct. Both Technicians are correct. A leaking heater core can cause a coolant smell to be present in the cab area. Some heater core leaks will cause coolant to leak onto the floorboard area of the truck. The HVAC drain tube will usually drain the leaking coolant near the firewall area.

    Answer D is incorrect. Both Technicians are correct.

4. A cycling clutch orifice tube (CCOT) A/C system is operating at 84° F (28.9° C) ambient temperature, the compressor clutch cycles several times per minute, and the suction line is warm. The high-side gauge shows lower than normal pressures. The most likely cause of this problem could be:

    A. Low refrigerant charge.

    B. Flooded evaporator.

    C. Restricted accumulator.

    D. Overcharge of refrigerant.

TASK A.3

    Answer A is correct. A low refrigerant charge would cause short cycling at the compressor as well as low system pressures.

    Answer B is incorrect. If the evaporator were flooded, then the suction line would be frosted due to the refrigerant still taking on heat.

    Answer C is incorrect. A restricted accumulator would cause the suction line to be frosted due to the restriction.

    Answer D is incorrect. A refrigerant overcharge would cause the high-side pressure to be higher than normal.

TASKS A.3,
D.8

5. While conducting a performance test on a semi-automatic HVAC system, a technician finds that only a small amount of air is directed to the windshield in defrost mode. The most likely cause of this problem is:

    A. A defective microprocessor.

    B. A defective blend door actuator.

    C. An open blower motor resistor.

    D. An improperly adjusted mode door cable.

    Answer A is incorrect. The microprocessor does not control the modes on a semi-automatic HVAC system. The processor controls only the temperature in these systems.

    Answer B is incorrect. A bad blend door actuator would only affect the temperature of the air.

    Answer C is incorrect. A blower resistor problem would only affect the blower speeds.

    Answer D is correct. An improperly adjusted mode door cable could cause the air to not be directed to the desired location.

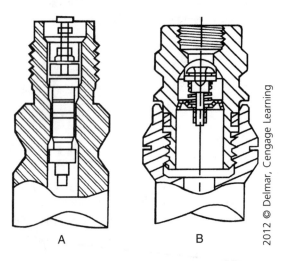

2012 © Delmar, Cengage Learning

TASK A.4

6. Referring to the figure, Technician A says that a system with service port fitting like figure A is designed to use R-134a as a refrigerant. Technician B says that the service port fitting represented by figure B is for use with R-12. Who is correct?

    A. A only

    B. B only

    C. Both A and B

    D. Neither A nor B

    Answer A is incorrect. Figure A is an R-12 service port fitting. The low-side fitting is 7/16 inch (11 mm) by 20 TPI and the high side is 3/8 inch (9.5 mm) by 24 TPI.

    Answer B is incorrect because figure B is an R-134a service port fitting. The low side fitting size is 13 mm (1/2 inch) and the high side fitting size is 16 mm (5/8 inch).

    Answer C is incorrect. Neither Technician is correct.

    Answer D is correct. Neither Technician is correct. Figure A is an R-12 fitting and figure B is an R-134a fitting.

7. Technician A says a TXV not regulating properly could cause reduced airflow from the instrument panel outlets. Technician B says a TXV not regulating properly could cause the evaporator to ice up. Who is correct?

TASK B.1.1

   A. A only

   B. B only

   C. Both A and B

   D. Neither A nor B

   Answer A is incorrect. Technician B is also correct.

   Answer B is incorrect. Technician A is also correct.

   Answer C is correct. Both Technicians are correct. A malfunctioning TXV could cause the evaporator core to ice up, thus causing reduced airflow to the panel outlets.

   Answer D is incorrect. Both Technicians are correct.

8. Technician A says that refrigerant with air contamination will have a higher static pressure than pure refrigerant. Technician B says that an electronic refrigerant identifier can be used to detect leak sealer additive. Who is correct?

TASK B.1.2

   A. A only

   B. B only

   C. Both A and B

   D. Neither A nor B

   Answer A is correct. Only Technician A is correct. Refrigerant that has air contamination will have a higher static pressure than pure refrigerant. A pressure/temperature chart can assist a technician in detecting air in a refrigerant system.

   Answer B is incorrect. Electronic refrigerant identifiers will not typically detect leak sealer additive. A refrigerant flow tool is needed to detect the leak sealer additive.

   Answer C is incorrect. Only Technician A is correct.

   Answer D is incorrect. Technician A is correct.

9. A band of frost on the A/C high-pressure liquid line at a point before the orifice tube indicates:

TASK B.3.2

   A. A defective compressor discharge valve.

   B. A restriction in the high-pressure hose.

   C. A clogged orifice tube.

   D. Moisture in the system.

   Answer A is incorrect. A faulty compressor discharge valve would not cause this problem. A faulty compressor discharge valve will cause poor cooling accompanied with unusual gauge pressures.

   Answer B is correct. A restriction in the high-pressure hose will result in a band of frost on the high-pressure line. All of the components in the high side of the system should be hot while the system is operating. A restriction anywhere in the high side will result in a cold spot or even frost to appear.

   Answer C is incorrect. A clogged orifice tube will not cause a band of frost to appear on the high-pressure hose. If the orifice tube becomes restricted, then the line downstream might be frosted.

   Answer D is incorrect. Moisture in the system could cause a restriction at the orifice tube and result in frost appearing at the orifice tube.

TASK B.1.6

10. When evacuating an A/C system, which manifold gauge hose is connected to the vacuum pump?

    A. The high-pressure hose

    B. The low-pressure hose

    C. The center service hose

    D. Any hose

Answer A is incorrect. The high-pressure hose is connected to the high-side service valve.

Answer B is incorrect. The low-pressure hose is connected to the low-side service valve.

Answer C is correct. When the system is completely empty, the technician connects the manifold gauge center hose to a vacuum pump. The center hose is typically the yellow hose in a manifold gauge set.

Answer D is incorrect. Connecting the incorrect hose to the vacuum pump can result in an inadequate evacuation.

TASK B.1.7

11. All of the following statements about nitrogen flushing the A/C system are true EXCEPT:

    A. The technician should install a pressure regulator on the supply tank.

    B. The technician should disconnect the A/C compressor.

    C. The technician should remove restrictive components (i.e., STV, TXV valve) from the system.

    D. Nitrogen must not be allowed to escape into the atmosphere.

Answer A is incorrect. To prevent damaging A/C system components, the nitrogen pressure must be regulated.

Answer B is incorrect. The A/C compressor must be disconnected before flushing the system to prevent debris from entering into it.

Answer C is incorrect. To avoid damaging restrictive components, they must be removed before flushing.

Answer D is correct. It is very acceptable to vent nitrogen to the atmosphere. The atmosphere already contains an abundant amount of nitrogen, so no harm will be done by venting nitrogen during A/C service.

TASK B.1.8

12. All of the following statements about charging an A/C system are true EXCEPT:

    A. Refrigerant may be installed through both service valves when the engine is not running.

    B. Refrigerant may be installed through the low-side service valve when the engine is running.

    C. Refrigerant may be installed through both service ports when the engine is running.

    D. You may install refrigerant directly from an approved charging station.

Answer A is incorrect. A technician can install refrigerant through both service valves when the engine is not running.

Answer B is incorrect. A technician can install refrigerant through the low-side service valve when the engine is running.

Answer C is correct. A technician should never attempt to charge an A/C system through the high-side service valve when the engine is running because this high pressure will likely exit the vehicle and enter the charging tank and could explode.

Answer D is incorrect. A technician can charge an A/C systems directly from an approved charging station.

13. During normal A/C operation, a loud hissing noise is heard and a cloud of vapor is discharged from under the vehicle. Technician A says that the excessive high-side pressure that caused the pressure relief valve on the A/C compressor to trip may have been the result of a defective engine cooling fan clutch. Technician B says that the relief valve may have tripped due to the refrigerant system being overcharged with refrigerant. Who is correct?

TASKS B.2.1, B.3.10, C.9

  A. A only

  B. B only

  C. Both A and B

  D. Neither A nor B

Answer A is incorrect. Technician B is also correct.

Answer B is incorrect. Technician A is also correct.

Answer C is correct. Both Technicians are correct. The excessive high-side pressure that caused the pressure relief valve on the A/C compressor to operate might have been the result of a faulty engine cooling fan clutch. A refrigerant overcharge could also cause the pressure relief valve to operate.

Answer D is incorrect. Both Technicians are correct.

14. A heavy truck with a faulty A/C pressure sensor is being diagnosed. Technician A says that this device is used to provide A/C pressure feedback to a control module. Technician B says that this device is typically mounted in the high side of the A/C system. Who is correct?

TASK B.2.2

  A. A only

  B. B only

  C. Both A and B

  D. Neither A nor B

Answer A is incorrect. Technician B is also correct.

Answer B is incorrect. Technician A is also correct.

Answer C is correct. Both Technicians are correct. A/C pressure sensors are often used on the high side of A/C systems to provide a voltage signal to a control module. This varying voltage is a more accurate method of controlling the A/C system. The range of the voltage is approximately 1 to 3 volts during operation.

Answer D is incorrect. Both Technicians are correct.

15. During a performance test, the A/C compressor clutch is observed to be slipping. Technician A says that the air gap was probably set improperly. Technician B says that the pressure plate needs to be resurfaced. Who is correct?

TASK B.2.4

  A. A only

  B. B only

  C. Both A and B

  D. Neither A nor B

Answer A is correct. Only Technician A is correct. If the air gap is too great, the clutch will slip. Compressor clutch air gap can be checked by sliding a feeler gauge between the drive and driven plates. If the air gap is above specs, shims can be removed to lower the gap.

Answer B is incorrect. The drive and driven plates must be replaced, not resurfaced.

Answer C is incorrect. Only Technician A is correct.

Answer D is incorrect. Technician A is correct.

TASK D.12

16. A truck has just received an HVAC repair. All of the following functions should be performed before delivering the truck back to the owner EXCEPT:

A. Inspect the fasteners for correct torque.

B. Recover the refrigerant.

C. Inspect the wiring connections.

D. Clear the diagnostic trouble codes.

Answer A is incorrect. It is advisable to inspect the fasteners that were installed during the repair to assure that they are tightened correctly.

Answer B is correct. The refrigerant would not need to be recovered after the repair has been made. Recovering the refrigerant is necessary when a refrigerant circuit component has to be serviced.

Answer C is incorrect. It is advisable to inspect the wiring connections that were handled during the repair. Each connection should be secure and tight to assure correct operation of the climate control system.

Answer D is incorrect. The diagnostic trouble codes should be cleared after the repair so that the truck returns to service without any evidence of past problems.

TASKS B.2.3, B.2.6

17. A knocking sound is coming from the area of the A/C compressor when in operation. When the compressor is shut off, the noise stops. The A/C system cools well and there are no indications of A/C system problems. Technician A says that the noise could be caused by a broken compressor mounting bracket. Technician B says that the noise could be caused by a restricted discharge line. Who is correct?

A. A only

B. B only

C. Both A and B

D. Neither A nor B

Answer A is correct. Only Technician A is correct. A broken compressor mounting bracket could produce a knocking noise when the compressor is operating.

Answer B is incorrect. A restricted discharge line would not be likely to produce a knocking noise when the compressor is operating. In addition, the system would not cool well if the discharge line were restricted.

Answer C is incorrect. Only Technician A is correct.

Answer D is incorrect. Technician A is correct.

TASK B.3.3

18. Any of the following can be used to clean road debris from the condenser fins EXCEPT:

A. A mild saline solution.

B. A soft whiskbroom.

C. Compressed air.

D. A mild soap and water solution.

Answer A is correct. A saline solution will cause corrosion.

Answer B is incorrect. A soft whiskbroom can be used to remove debris from the condenser fins.

Answer C is incorrect. Compressed air can be used to remove debris from the condenser fins.

Answer D is incorrect. A soap and water solution can be used to remove debris from the condenser fins.

19. In servicing the expansion valve, which of the following can a technician perform?

    A. Adjust the expansion valve using a torque wrench.
    B. Adjust the expansion valve using an Allen wrench.
    C. Adjust the expansion valve using a screwdriver.
    D. The expansion valve cannot be adjusted.

TASK B.3.6

    Answer A is incorrect. You cannot adjust an expansion valve with any process.

    Answer B is incorrect. You cannot adjust an expansion valve with any process.

    Answer C is incorrect. You cannot adjust an expansion valve with any process.

    Answer D is correct. The expansion valve is not an adjustable device.

20. Approximately how much refrigerant oil must be added to a newly replaced evaporator core?

    A. None
    B. 3 ounces
    C. 9 ounces
    D. 14.5 ounces

TASK B.3.8

    Answer A is incorrect. Refrigerant oil must be added to the evaporator prior to installation.

    Answer B is correct. Most manufacturers recommend that 3 ounces of oil be added to a new evaporator.

    Answer C is incorrect. Adding 9 ounces of refrigerant oil would cause an excessive system oil level.

    Answer D is incorrect. The entire A/C system is likely to contain about 14.5 ounces of oil.

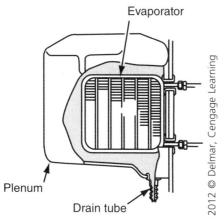

Evaporator

Plenum

Drain tube

2012 © Delmar, Cengage Learning

21. Referring to the figure, which of the following conditions would most likely occur if the drain tube becomes blocked?

    A. The A/C system would become colder.
    B. The heater core would become restricted.
    C. Water could leak onto the floorboard.
    D. The evaporator core would leak.

TASK B.3.9

    Answer A is incorrect. A blocked A/C drain tube would not improve A/C performance. The driver may experience a mist of water being directed out of the vents, however.

    Answer B is incorrect. A blocked A/C drain tube would not cause the heater core to become restricted.

    Answer C is correct. A blocked A/C drain tube could cause water to leak onto the floorboard of the truck due to overfilling of the HVAC duct. A technician can sometimes clear a blocked drain tube by probing the opening with a piece of stiff wire.

    Answer D is incorrect. A blocked A/C drain tube would not cause the evaporator core to leak. The driver may, however, experience a water mist being directed out of the vents with the blower motor on the higher speeds.

TASK B.3.10

22. The A/C compressor high-pressure relief valve:

A.   Is calibrated by shimming it to the proper depth.

B.   Must be replaced if it ever vents refrigerant from the system.

C.   Will reset itself when A/C system pressure returns to a safe level.

D.   Is not used in R-134a systems.

Answer A is incorrect. The relief valve cannot be calibrated.

Answer B is incorrect. The relief valve does not need to be replaced if it vents refrigerant to the atmosphere and then resets itself.

Answer C is correct. The valve will reset itself when A/C system pressure returns to a safe level. The valve is designed to release pressure at 450–550 psi.

Answer D is incorrect. All mobile A/C systems use a high-pressure relief valve.

TASK C.2

23. Which of the following is LEAST LIKELY to cause windshield fogging in the DEFROST mode?

A.   A leaking heater core

B.   A clogged evaporator drain

C.   A water leak into the plenum chamber

D.   Moisture in the refrigerant

Answer A is incorrect. A leaking heater core could cause windshield fogging in the DEFROST mode.

Answer B is incorrect. A clogged evaporator drain could cause windshield fogging in the DEFROST mode.

Answer C is incorrect. An exterior water leak into the air intake plenum could cause windshield fogging in the DEFROST mode.

Answer D is correct. Moisture in the refrigerant could cause acid to form when it mixes with the refrigerant or it could cause ice to form near the expansion device.

TASK C.4

24. The upper radiator hose has a slight bulge. Technician A says that the hose does not need to be replaced unless it appears to be cracked. Technician B says that the bulge indicates a weak spot and the hose should be replaced. Who is correct?

A.   A only

B.   B only

C.   Both A and B

D.   Neither A nor B

Answer A is incorrect. A bulge in the hose indicates a weak spot.

Answer B is correct. Only Technician B is correct. The hose should be replaced immediately. All of the coolant hoses should be thoroughly checked at this time. If the hoses are the same age as the failed hose, it is advisable to replace all of them at the same time.

Answer C is incorrect. Only Technician B is correct.

Answer D is incorrect. Technician B is correct.

25. A scan tool can be used for all of the following functions on a climate control computer EXCEPT:

    A. Calibrating the blower resistor.

    B. Retrieving trouble codes.

    C. Displaying sensor data.

    D. Displaying switch data.

TASK D.11

Answer A is correct. A scan tool is not used to calibrate a blower resistor. A blower resistor does not have any logic functions that need to be calibrated.

Answer B is incorrect. A scan tool can be used to retrieve trouble codes from a climate control computer. These codes can be used to assist the technician in troubleshooting problems in the climate control system.

Answer C is incorrect. A scan tool can be used to display sensor data for the technician. The sensor data will often reveal problems in the climate control system.

Answer D is incorrect. A scan tool can be used to display switch data for the technician. The switch data will often reveal problems in the climate control system.

26. All of these statements about cooling system service are true EXCEPT:

    A. When the cooling system pressure is increased, the boiling point increases.

    B. The boiling point decreases when more antifreeze is added to the coolant.

    C. Good quality ethylene glycol antifreeze contains antirust inhibitors.

    D. Coolant solutions should be recovered, recycled, and handled as hazardous waste.

TASK C.8

Answer A is incorrect. When the cooling system pressure is increased, the boiling point increases. For every 1 psi of pressure that is added to the cooling system, the boiling point is raised 3° F.

Answer B is correct. When more antifreeze is added to the coolant, the boiling point increases.

Answer C is incorrect. A good quality ethylene glycol antifreeze contains antirust inhibitors.

Answer D is incorrect. Coolant solutions must be recovered, recycled, and handled as hazardous waste.

27. When the engine cooling fan viscous clutch is disengaged:

    A. The fan blade will remain stationary.

    B. The engine idle will drop.

    C. The radiator shutters must be closed.

    D. The fan blade may freewheel at a reduced speed.

TASK C.9

Answer A is incorrect. The fan blade will freewheel at a reduced speed due to friction/viscous forces.

Answer B is incorrect. The engine idle will rise slightly or be unaffected.

Answer C is incorrect. The shutters may or may not be closed.

Answer D is correct. The fan blade may freewheel at a reduced speed. A viscous fan clutch operates by not requiring the fan to turn at engine speed when the temperature is cool. As the fan heats up, the viscous connection speeds up the fan to help cool the engine back down.

TASK C.10

28. Technician A says that a vacuum-operated coolant control valve can be tested with a scan tool. Technician B says a vacuum-operated coolant control valve can be tested by applying and releasing vacuum and checking if the valve arm moves freely both ways. Who is correct?

    A. A only

    B. B only

    C. Both A and B

    D. Neither A nor B

Answer A is incorrect. Vacuum-operated coolant control valves are mechanical devices that do not have electronic connections. A scan tool would not be capable of communicating with this type of valve.

Answer B is correct. Only Technician B is correct. This process allows the technician to check the specific coolant flow valve operation. The valve should hold the vacuum for at least one minute without leaking down.

Answer C is incorrect. Only Technician B is correct.

Answer D is incorrect. Technician B is correct.

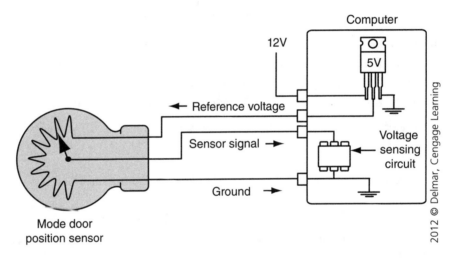

29. Referring to the figure above, which of the following readings would most likely be present on the sensor signal wire with the wiper in the current position?

TASK D.1

    A. 0.0 volts

    B. 0.5 volts

    C. 2.5 volts

    D. 4.5 volts

Answer A is incorrect. The sensor signal would never read 0.0 volts unless the wiper broke off.

Answer B is incorrect. The sensor signal would be approximately 0.5 volts when the wiper is near the bottom of the resistive strip.

Answer C is incorrect. The sensor signal would be approximately 2.5 volts when the wiper is near the middle of the resistive strip.

Answer D is correct. The signal voltage would be approximately 4.5 volts since the wiper is up near the top of the resistive strip.

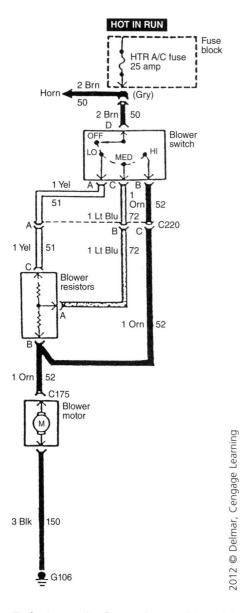

30. Referring to the figure above, which of the following statements is LEAST LIKELY to be correct concerning the blower circuit?

     A. The blower switch receives power from connector C220.

     B. The blower motor would only have high speed if the blower resistor were removed.

     TASK D.2

     C. The blower motor is grounded at G106.

     D. The blower circuit has three speeds.

Answer A is correct. The blower switch does not receive power from connector C220. The blower switch receives power from the heater/air conditioning fuse at the fuse block.

Answer B is incorrect. The blower motor would have only high speed if the blower resistor were removed. Power does not enter the blower resistor when the blower switch is in the high position.

Answer C is incorrect. G106 provides ground for the blower motor. The blower speeds are controlled on the power side of the circuit.

Answer D is incorrect. The blower motor has low speed, medium speed, and high speed.

TASK D.3

31. On accumulator-type systems with the compressor cycling switch located on the accumulator, the switch senses:

   A. Outside temperature.

   B. Accumulator pressure.

   C. Accumulator temperature.

   D. Engine compartment temperature.

   Answer A is incorrect. The cycling switch does not sense outside temperature.

   Answer B is correct. The cycling switch is mounted in the accumulator where it senses pressure. The cycling switch usually opens at about 20–25 psi and closes at about 40–45 psi.

   Answer C is incorrect. The cycling switch senses accumulator pressure, not temperature.

   Answer D is incorrect. The cycling switch does not sense engine ambient temperature.

TASK D.4

32. Technician A says that the HVAC systems on most medium-duty trucks with diesel engines operate totally independently from the engine control system. Technician B says that most of the HVAC systems on these vehicles provide inputs and receive outputs from the engine control unit. Who is correct?

   A. A only

   B. B only

   C. Both A and B

   D. Neither A nor B

   Answer A is incorrect. Most late-model A/C systems are controlled by the engine management system.

   Answer B is correct. Only Technician B is correct. Most late-model A/C systems are controlled by the engine control system.

   Answer C is incorrect. Only Technician B is correct.

   Answer D is incorrect. Technician B is correct.

TASK D.5

33. A truck with an air-controlled fan clutch is being diagnosed for a fan problem. Technician A says that some trucks have a toggle switch on the dash to enable the fan. Technician B says that the fan solenoid controls the air supply to the fan clutch. Who is correct?

   A. A only

   B. B only

   C. Both A and B

   D. Neither A nor B

   Answer A is incorrect. Technician B is also correct.

   Answer B is incorrect. Technician A is also correct.

   Answer C is correct. Both Technicians are correct. A dash-mounted toggle switch to manually engage the fan is a common option for heavy trucks. The fan solenoid controls the air supply to the fan clutch. The fan solenoid receives a signal from the engine computer or the manual fan switch.

   Answer D is incorrect. Both Technicians are correct.

34. A driver complains that the air-operated heater outputs hot air when the temperature control lever is in the COLD position. Which of the following conditions would be the LEAST LIKELY cause of this problem?

TASKS D.7,
D.9

  A. A defective coolant control valve

  B. A defective engine coolant temperature sensor

  C. A defective air control solenoid

  D. A defective blend air door control cylinder

  Answer A is incorrect. A defective coolant control valve could cause this problem by not regulating coolant flow into the heater core properly.

  Answer B is correct. A defective engine coolant temperature sensor should not cause a problem with the temperature control system.

  Answer C is incorrect. A defective air control solenoid could cause this problem by not changing the blend door in the correct way.

  Answer D is incorrect. A defective blend air door control cylinder could cause this problem.

35. Inadequate airflow from one vent could be caused by which of the following conditions?

TASK D.8

  A. Misaligned air duct

  B. Faulty blower resistor

  C. Clogged heater core

  D. High ambient humidity

  Answer A is correct. A misaligned air duct could cause inadequate airflow from one or more vents. This problem could also be caused by a foreign object blocking the duct.

  Answer B is incorrect. A faulty blower resistor will affect airflow from all vents in one or more blower speed settings.

  Answer C is incorrect. A clogged heater core will not affect output airflow. A clogged core would cause a lack of heat problem.

  Answer D is incorrect. High humidity will not affect output airflow.

36. Which of the following procedures would be the LEAST LIKELY one performed on a truck that has just been repaired by a technician?

TASK D.12

  A. Inspect the components for leaks.

  B. Clear the diagnostic trouble codes.

  C. Identify the refrigerant.

  D. Operate the system to monitor for correct operation.

  Answer A is incorrect. It is advisable to inspect the components of the climate control system for leaks prior to returning the truck to the owner.

  Answer B is incorrect. The diagnostic trouble codes should be cleared after the repair so that the truck returns to service without any evidence of past problems.

  Answer C is correct. This step should be performed at the beginning of the repair process. The technician should identify the refrigerant prior to connecting any manifold gauges or recovery equipment to the vehicle.

  Answer D is incorrect. It is advisable to operate the system to make sure it is performing well before returning the truck to the owner.

TASK B.1.12

37. Owners of approved refrigerant recycling equipment must maintain records that demonstrate that:

A. Only certified technicians operate the equipment.
B. The equipment is operated under the supervision of certified technicians.
C. Technicians who are undergoing certification training operate the equipment.
D. The equipment is operated only when a certified technician is on the premises.

Answer A is correct. The Clean Air Act (CAA) established the rule of certification. This rule states that all persons authorized to operate the equipment must be certified under the Act.

Answer B is incorrect. This would be in violation of the CAA.

Answer C is incorrect. This would be in violation of the CAA.

Answer D is incorrect. This would be in violation of the CAA.

TASK B.1.5

38. To prevent overfilling recovery cylinders, the service technician must:

A. Monitor cylinder pressure as the cylinder is being filled.
B. Monitor cylinder weight as the cylinder is being filled.
C. Make sure cylinder safety relief valves are in place and operational.
D. Occasionally shake the cylinder and observe any change of pressure while filling.

Answer A is incorrect. Cylinder pressure is not an accurate measure of the content level.

Answer B is correct. The total weight of the cylinder must not exceed the weight of the cylinder when it is empty plus the maximum rated net weight.

Answer C is incorrect. Refrigerant cylinders are not equipped with safety relief valves.

Answer D is incorrect. Shaking the cylinder is not an accurate measure of contents level.

TASK D.12

39. A truck A/C system has just been repaired. Technician A says that the system should be overcharged by 1/2 pound to account for future leaks in the refrigerant system. Technician B says that the system should be operated to make sure that it functions correctly. Who is correct?

A. A only
B. B only
C. Both A and B
D. Neither A nor B

Answer A is incorrect. It is never advisable to intentionally overcharge a refrigerant system. This practice will cause decreased system performance and possible early component failure.

Answer B is correct. Only Technician B is correct. It is a good practice to operate the A/C system after a repair has been made to guarantee that it is performing up to the normal expected level.

Answer C is incorrect. Only Technician B is correct.

Answer D is incorrect. Technician B is correct.

TASK B.1.10

40. Before disposing of an empty or nearly empty original container that was used to ship refrigerant from the factory, which of the following should a technician do?

A. Clean the container and keep it for storage of recycled refrigerant.
B. Open the valve completely and paint an X on the cylinder.
C. Flush the container with oil and nitrogen to keep it from rusting.
D. Recover remaining refrigerant, evacuate the cylinder, and mark it empty.

Answer A is incorrect. Original refrigerant containers should not be used to store recycled refrigerant.

Answer B is incorrect. If the valve is simply opened, any remaining refrigerant will be vented to the atmosphere.

Answer C is incorrect. There is no reason to introduce oil into the cylinder.

Answer D is correct. After any remaining refrigerant is recovered, the cylinder should be evacuated, marked, and recycled for scrap metal.

PREPARATION EXAM 4—ANSWER KEY

1.	D	21.	A
2.	A	22.	B
3.	B	23.	D
4.	C	24.	A
5.	D	25.	B
6.	C	26.	B
7.	A	27.	B
8.	C	28.	D
9.	B	29.	A
10.	C	30.	B
11.	D	31.	B
12.	C	32.	A
13.	B	33.	D
14.	C	34.	C
15.	C	35.	A
16.	A	36.	A
17.	B	37.	B
18.	D	38.	B
19.	D	39.	B
20.	C	40.	A

PREPARATION EXAM 4—EXPLANATIONS

1. Which of the following actions would be LEAST LIKELY to happen when a technician receives a repair order on a truck with an A/C complaint?

   A. Verify the complaint.

   B. Road test the truck.

   C. Review past maintenance records.

   D. Perform a heat load test on the refrigerant system.

TASK A1

Answer A is incorrect. The technician should verify the complaint soon after receiving the repair order. Verifying the complaint will prove that there is an active problem that can be diagnosed.

Answer B is incorrect. The technician will often road test the truck to attempt to simulate the conditions the driver is complaining about.

Answer C is incorrect. The technician will sometimes review the maintenance records of the truck to identify possible systems that may need some attention.

Answer D is correct. The technician would not run a heat load test right after receiving a repair order for an A/C system complaint on a truck. Several other steps will take place before running an intricate heat load test.

TASK A.2

2. A hissing sound is heard under the hood area of a truck after the engine has been turned off. The noise stops after about one minute and only happens when the air conditioner or defroster has been used. Technician A says that the A/C system is equalizing. Technician B says that the noise is not normal and the truck should be towed to the nearest repair shop. Who is correct?

    A. A only
    B. B only
    C. Both A and B
    D. Neither A nor B

    Answer A is correct. Only Technician A is correct. A hissing noise is a normal occurrence for a truck equipped with an A/C system. The high-side and low-side pressures will equalize after the engine is turned off.

    Answer B is incorrect. The equalizing noise in the A/C system is a normal sound and no cause for concern.

    Answer C is incorrect. Only Technician A is correct.

    Answer D is incorrect. Technician A is correct.

TASK A.2

3. A compressor has excessive air gap at the clutch drive plate. Technician A says that the clutch could squeal when the compressor disengages. Technician B says that the clutch could squeal when the ambient temperature is very high. Who is correct?

    A. A only
    B. B only
    C. Both A and B
    D. Neither A nor B

    Answer A is incorrect. Excessive air gap at the A/C clutch drive plate would not cause a squealing noise when the compressor disengages. This problem would likely cause noise when the compressor engages.

    Answer B is correct. Only Technician B is correct. Excessive air gap at the A/C clutch drive plate could cause the clutch plate to squeal on very hot days due to the increased high-side pressures that would be present.

    Answer C is incorrect. Only Technician B is correct.

    Answer D is incorrect. Technician B is correct.

TASK A.3

4. During an HVAC performance test, the Technician notices that the A/C compressor outlet is nearly as hot as the upper radiator hose. Technician A says that this is a normal condition. Technician B says that the system is overcharged. Who is correct?

    A. A only
    B. B only
    C. Both A and B
    D. Neither A nor B

    Answer A is correct. Only Technician A is correct. The high-pressure refrigerant at the compressor outlet can raise the temperature of the outlet to that of the engine coolant or even higher on a hot and humid day.

    Answer B is incorrect. This is a normal condition caused by the compressor raising the pressure/temperature of the vapor refrigerant.

    Answer C is incorrect. Only Technician A is correct.

    Answer D is incorrect. Technician A is correct.

5. Technician A says that the compressor clutch should not make an audible sound when engaging and disengaging. Technician B says that ice forming on the suction line is a sign of a properly functioning A/C system. Who is correct?

TASK A.3

A. A only

B. B only

C. Both A and B

D. Neither A nor B

Answer A is incorrect. A cycling compressor is typically loud enough to be heard by a technician.

Answer B is incorrect. Ice that is present at any point of the A/C system is a sign that something is not working correctly.

Answer C is incorrect. Neither Technician is correct.

Answer D is correct. Neither Technician is correct. A Technician should normally be able to hear the compressor cycling on and off. A properly functioning A/C should never have ice forming on components.

6. Technician A states that systems that use an orifice tube use an accumulator. Technician B states that systems that use a TXV use a receiver/drier. Who is correct?

TASK A.4

A. A only

B. B only

C. Both A and B

D. Neither A nor B

Answer A is incorrect. Technician B is also correct.

Answer B is incorrect. Technician A is also correct.

Answer C is correct. Both Technicians are correct. Orifice tube systems use an accumulator in the suction line to store and dry the refrigerant. TXV systems use a receiver/dryer in the liquid line to store and dry the refrigerant.

Answer D is incorrect. Both Technicians are correct.

7. In MAX A/C mode:

TASK B.1.1

A. The outside air door is closed.

B. The defroster door is open.

C. The compressor clutch cannot disengage.

D. The blower is disabled.

Answer A is correct. In the MAX A/C mode the outside air door is closed. Running the A/C in the MAX mode will allow the air inside the cab to re-circulate within the cab, resulting in colder duct temperatures.

Answer B is incorrect. The defroster door is closed in the MAX A/C mode.

Answer C is incorrect. The compressor clutch will cycle normally in the MAX A/C mode.

Answer D is incorrect. The blower functions normally in the MAX A/C mode.

TASK B.1.3

8. An A/C system with excessive high-side pressure could be the result of all of the following EXCEPT:

    A. An overcharge of refrigerant.
    B. An overheated engine.
    C. Restricted airflow through the condenser.
    D. Ice buildup on the orifice tube screen.

    Answer A is incorrect. A refrigerant overcharge will cause high system pressures.

    Answer B is incorrect. An overheated engine will cause elevated high-side pressure due to high condenser temperature.

    Answer C is incorrect. A restricted airflow through the condenser will cause elevated high-side pressure due to the lack of heat transfer resulting from restricted airflow.

    Answer D is correct. A blockage of the orifice tube screen will cause low, high-side pressure.

TASK B.3.2

9. Oil and dirt accumulation on an A/C hose connection may indicate:

    A. Excessive pressure in the system.
    B. A refrigerant leak.
    C. A defective compressor shaft seal.
    D. That there is too much oil in the system.

    Answer A is incorrect. Excessive pressure in the system would have to be high enough to rupture a hose, and that is unlikely.

    Answer B is correct. Refrigerant leaking from a hose connection will usually contain some refrigeration oil, which quickly collects dirt.

    Answer C is incorrect. A shaft seal will only cause an oil residue at the front of the compressor.

    Answer D is incorrect. Too much oil in the system reduces system efficiency and causes low cooling because the oil will coat the heat exchangers.

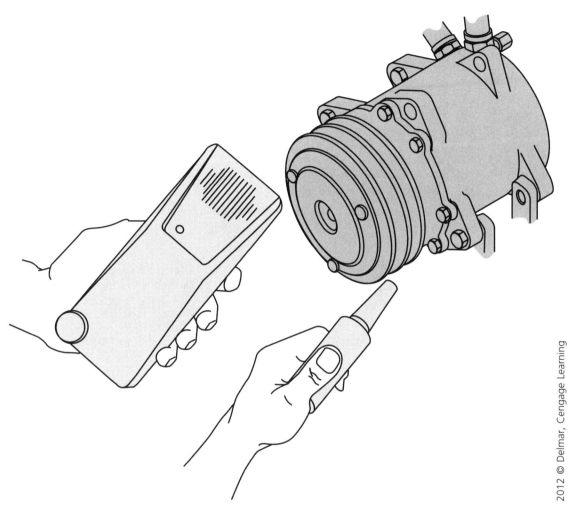

2012 © Delmar, Cengage Learning

10.   What tool is the technician using in the figure above?

A.   Refrigerant identifier

B.   Belt tension gauge

C.   Refrigerant leak detector

D.   Refrigerant electronic manifold

TASK B.1.5

Answer A is incorrect. The tool is not a refrigerant identifier. An identifier has a hose connection that connects to the service fitting on the A/C system.

Answer B is incorrect. The tool is not a belt tension gauge. A belt tension gauge is used to determine the tightness of the drive belt.

Answer C is correct. The tool in the figure is a refrigerant leak detector. This tool provides an audible sound when it senses refrigerant.

Answer D is incorrect. The tool is not an electronic manifold device. Electronic manifold sets would have the normal service hoses that analog manifolds have always used.

TASK B.1.7

11. Technician A says some manufacturers recommend the installation of an in-line filter between the evaporator and the compressor as an alternative to refrigerant system flushing. Technician B says an in-line filter containing a fixed orifice may be installed and the original orifice tube left in the system. Who is correct?

 A. A only

 B. B only

 C. Both A and B

 D. Neither A nor B

Answer A is incorrect. In-line filters are not a substitute for flushing or replacing components that have been contaminated by debris.

Answer B is incorrect. If an in-line filter contains an orifice, then the original orifice must be removed.

Answer C is incorrect. Neither Technician is correct.

Answer D correct. Neither Technician is correct. In-line filters are sometimes recommended after flushing and/or component replacement as a method to prevent debris from entering into the compressor. If an in-line filter is installed that contains an orifice tube, then the original orifice tube must be removed.

TASK B.1.8

12. Which of these statements is correct concerning the method to recharge a heavy truck A/C system?

 A. The charging process is complete when the system reaches the correct evaporator temperature.

 B. When the low side no longer moves from a vacuum to a pressure, the process is complete.

 C. The truck engine must be running during recharging.

 D. Either a high-side or a low-side charging process can be used.

Answer A is incorrect. A technician completes the charging process when the correct weight of refrigerant has entered the system.

Answer B is incorrect. If the low side does not move from a vacuum to a pressure, there is a restriction.

Answer C is incorrect. The truck engine does not need to be running when recharging with a recharging machine.

Answer D is correct. Original equipment manufacturers (OEMs) recommend either a high-side (liquid) or a low-side (vapor) charging process.

TASK B.2.2

13. The A/C high-pressure switch is used to:

 A. Boost the system high-side pressure.

 B. Open the circuit to the A/C compressor clutch coil when the high-side pressure reaches its upper limit.

 C. Ensure that system pressure remains at the upper limit.

 D. Vent refrigerant from the compressor in the event of extremely high system pressure.

Answer A is incorrect. The high-pressure switch does not provide a boosting function.

Answer B is correct. The A/C high-pressure switch opens the electrical circuit to the compressor clutch coil when high-side pressure reaches its upper limit. The high-pressure switch typically opens at about 380–420 psi.

Answer C is incorrect. The high-pressure switch does not maintain pressure.

Answer D is incorrect. The high-pressure switch does not perform a venting function.

14. A serpentine belt is being replaced on a diesel-powered truck and the tensioner will not snap back after being released. Technician A says that the tensioner spring could be broken and the tensioner will need to be replaced. Technician B says that the idler pulley is jammed and will need to be replaced. Who is correct?

TASK B.2.3

A. A only

B. B only

C. Both A and B

D. Neither A nor B

Answer A is correct. Only Technician A is correct. The tensioner will need to be replaced if it does not snap back after it is released.

Answer B is incorrect. The idler pulley is a separate device from the tensioner pulley.

Answer C is incorrect. Only Technician A is correct.

Answer D is incorrect. Technician A is correct.

15. To check and adjust the A/C compressor lubricant level:

TASK B.2.5

A. Quickly purge the system and add oil charges to refill it.

B. Open the drain plug and crank the engine until the compressor is empty, then pump fresh oil into the compressor.

C. Remove the compressor from the vehicle, drain the oil, and add the specified quantity of fresh oil.

D. Add refrigerant oil until you can see the oil level.

Answer A is incorrect. This is not an acceptable method of adjusting the lubricant level in the system.

Answer B is incorrect. This is not an acceptable method because the A/C system must first be recovered and the compressor removed from the vehicle.

Answer C is correct. The only accurate way to measure the amount of oil in a compressor is to remove the compressor and drain the oil and measure. If less than two ounces is removed from the compressor, then add two ounces of new oil to the compressor. If more than two ounces was removed, then add the amount removed of new oil. Always follow the compressor manufacturer recommendations in regard to adding oil to the compressor.

Answer D is incorrect. This method will result in overfilling the compressor.

2012 © Delmar, Cengage Learning

TASK B.2.5

16. Referring to the figure above, Technician A says that the old compressor oil should be drained and measured when replacing an A/C compressor. Technician B says that the old oil should be added to the new compressor. Who is correct?

   A. A only

   B. B only

   C. Both A and B

   D. Neither A nor B

Answer A is correct. Only Technician A is correct. The oil should be drained from the old compressor whenever a replacement compressor is to be installed. If more than two ounces of oil were drained from the old compressor, then that amount should be added to the new compressor. If less than two ounces of oil were drained from the old compressor, then two ounces of new oil should be added to the new compressor.

Answer B is incorrect. Used oil should never be put back into an A/C component or system. New oil should be added to the new compressor (unless stated by the compressor manufacturer).

Answer C is incorrect. Only Technician A is correct.

Answer D is incorrect. Technician A is correct.

17. A cracked A/C compressor mounting plate could cause all of the following symptoms EXCEPT:

    A. Drive belt wear.

    B. Internal compressor damage.

    C. Vibration with the A/C compressor clutch engaged.

    D. Drive belt squeal or chatter.

TASK B.2.6

Answer A is incorrect. A cracked mounting plate could cause drive belt wear by not holding the compressor in place. This would misalign the compressor and cause the belt to wear.

Answer B is correct. A cracked mounting plate will not cause internal compressor damage.

Answer C is incorrect. A cracked mounting plate can cause a vibration with the compressor clutch engaged. This vibration can appear to be caused by a faulty compressor but is just a cracked mounting plate.

Answer D is incorrect. A cracked mounting plate can cause belt squeal or chatter because of misaligned pulleys.

18. Technician A says that one ounce of refrigerant oil should be added to the accumulator/drier prior to installing it on a truck. Technician B says that five ounces of oil should be added to the suction line prior to installing it on a truck. Who is correct?

    A. A only

    B. B only

    C. Both A and B

    D. Neither A nor B

TASK B.3.1

Answer A is correct. Only Technician A is correct. It is a good practice to add at least one ounce of refrigerant oil to the accumulator/drier prior to installing it on a truck. The Technician should follow the manufacturer's recommendations when performing this repair.

Answer B is incorrect. Adding five ounces of oil to the suction line would not be a normal practice. This would likely cause compressor problems due to the large quantity of oil entering at one time.

Answer C is incorrect. Only Technician A is correct.

Answer D is incorrect. Technician A is correct.

19. Which of the following should be used to lubricate replacement A/C hose o-rings?

    A. Petroleum jelly

    B. Transmission fluid

    C. Silicone grease

    D. Refrigerant oil

TASK B.3.2

Answer A is incorrect. Petroleum jelly must not be used to lubricate o-rings.

Answer B is incorrect. Transmission fluid can damage A/C o-rings.

Answer C is incorrect. Silicone grease must not be introduced into the system.

Answer D is correct. Mineral-based refrigerant oil is the recommended lubricant for A/C o-rings.

TASK B.3.4

20. Technician A says that deformed or improperly aligned condenser mounting insulators will damage the A/C compressor. Technician B says that deformed or improperly aligned condenser mounting insulators could damage the condenser and refrigerant lines. Who is correct?

A. A only

B. B only

C. Both A and B

D. Neither A nor B

Answer A is incorrect. Damaged or misaligned condenser mounts should not affect compressor operation.

Answer B is correct. Only Technician B is correct. Deformed or improperly aligned condenser mounting insulators could damage the condenser and refrigerant lines. The insulators are made from rubber and act to cushion the vibration of the truck and engine from the condenser and lines. If the insulators are deformed or misaligned, leaks could develop in these areas.

Answer C is incorrect. Only Technician B is correct.

Answer D is incorrect. Technician B is correct.

TASK B.3.6

21. The thermal bulb on an expansion valve must be installed in contact with:

A. The evaporator outlet tube.

B. The condenser fins.

C. The suction hose.

D. The refrigerant.

Answer A is correct. The thermal bulb must be in contact with the evaporator outlet tube. The thermal bulb serves as the temperature sensing part of the TXV. If the bulb senses warm temperatures, the TXV opens to allow more refrigerant to enter the evaporator core. If the bulb senses cold temperatures, the TXV closes to restrict the flow of refrigerant into the evaporator core.

Answer B is incorrect. This area is the incorrect location. The condenser is in the high side of the system.

Answer C is incorrect. This is close to where the thermal bulb is located, but the bulb is inside the HVAC duct attached to the evaporator outlet line.

Answer D is incorrect. The thermal bulb does not directly contact the refrigerant.

TASK B.3.9

22. A whistling noise coming from under the passenger side dash with the blower motor on high speed might indicate:

A. A clogged evaporator drain.

B. A cracked evaporator case.

C. A broken blend door cable.

D. A low refrigerant charge.

Answer A is incorrect. A clogged evaporator drain will cause water to leak into the cab or possibly a mildew smell.

Answer B is correct. A cracked evaporator case could cause a whistling noise with the blower motor on high speed.

Answer C is incorrect. A broken blend door cable will cause a loss of control of the temperature of the discharge air from the HVAC system.

Answer D is incorrect. A low refrigerant charge will cause poor cooling and low system pressures.

23. A heater does not supply the cab with enough heat. The coolant level and blower test OK. Technician A says an improperly adjusted temperature control cable could be the cause. Technician B says a clogged heater core could be the cause. Who is correct?

    TASK C.1

    A. A only
    B. B only
    C. Both A and B
    D. Neither A nor B

    Answer A is incorrect. Technician B is also correct.

    Answer B is incorrect. Technician A is also correct.

    Answer C is correct. Both technicians are correct. A low heat problem could be caused by a misadjusted temperature control cable or by a clogged heater core.

    Answer D is incorrect. Both Technicians are correct.

24. Coolant conditioner performs all of the following tasks EXCEPT:

    A. It raises the boiling point of the coolant.
    B. It inhibits rust and debris from the coolant.
    C. It lubricates the cooling system internally.
    D. It prevents cavitation corrosion of wet cylinder liners.

    TASK C.3

    Answer A is correct. Antifreeze raises the boiling point of the coolant, not the conditioner.

    Answer B is incorrect. The coolant conditioner cartridge filters particulate from the coolant.

    Answer C is incorrect. The coolant conditioner internally lubricates the cooling system.

    Answer D is incorrect. The coolant conditioner prevents cavitation corrosion of the cylinder liners.

25. A coiled spring inside a radiator hose is used to:

    A. Pre-form the hose.
    B. Prevent the hose from collapsing.
    C. Increase the hose burst pressure.
    D. Eliminate cavitation.

    TASK C.4

    Answer A is incorrect. The shape of a preformed hose is determined during the manufacturing process.

    Answer B is correct. The spring provides internal support for the lower hose so it will not collapse as the water pump pulls water from the radiator into the engine.

    Answer C is incorrect. The pressure resilience of the hose comes from the rubber.

    Answer D is incorrect. A spring inside of a hose does not prevent cavitation, which is caused when air bubbles form and collapse.

TASK C.5

26. A cooling system pressure tester can be used to test:

    A. Thermostats.
    B. Radiators, pressure caps, and hoses.
    C. A/C leaks.
    D. The blend door actuator diaphragm.

    Answer A is incorrect. Thermostats can be tested by monitoring engine temperature during the warm-up process to make sure the thermostat does not open too early. The thermostat can also be tested by removing it from the truck and submerging it in hot water to see when it opens.

    Answer B is correct. A technician uses a pressure tester to test radiators, pressure caps, and hoses for leaks. Caution should be taken to not apply pressure higher than the radiator cap is rated.

    Answer C is incorrect. A/C leaks must be located using a leak detector.

    Answer D is incorrect. Vacuum diaphragms are tested using a hand-held vacuum pump.

TASKS C.6, C.7

27. The LEAST LIKELY cause of poor coolant circulation in a truck with a down-flow radiator is:

    A. A defective thermostat.
    B. An eroded water pump impeller.
    C. A collapsed upper radiator hose.
    D. A collapsed lower radiator hose.

    Answer A is incorrect. A defective thermostat can cause poor coolant circulation.

    Answer B is incorrect. An eroded water pump impeller will cause poor coolant circulation.

    Answer C is correct. The upper hose is usually under pressure and is unlikely to collapse.

    Answer D is incorrect. The lower hose is on the suction side of the water pump and could collapse, even though it usually contains a spring to keep this from happening.

TASK C.9

28. All of the following statements about the chassis air system are true EXCEPT:

    A. It is important to keep water drained from the system.
    B. System air can be used to operate the fan clutch.
    C. System air can be used to operate the radiator shutters.
    D. System air actuates the engine thermostat.

    Answer A is incorrect. It is important to keep water drained from the system to reduce internal corrosion.

    Answer B is incorrect. Some fan clutches are operated using chassis air.

    Answer C is incorrect. Most radiator shutter systems are air-operated.

    Answer D is correct. System air is not used to actuate the engine thermostat. Thermostats are self contained and internally open and close at their rated temperature to maintain the correct engine temperature.

TASK D.1.1

29. A blown HVAC system fuse could indicate any of the following EXCEPT:

    A. A short circuit to ground in the blower circuit.
    B. A short circuit to ground in the blend door actuator.
    C. A short circuit in the engine coolant temperature sensor (ECT).
    D. A damaged wiring harness connector.

    Answer A is incorrect. A short circuit in the blower circuit could cause a blown fuse.

    Answer B is incorrect. A short circuit in an actuator motor could cause a blown fuse.

    Answer C is correct A shorted ECT sensor will cause a code to set, but will not blow a fuse.

    Answer D is incorrect. A damaged connector could cause a short circuit and blow a fuse.

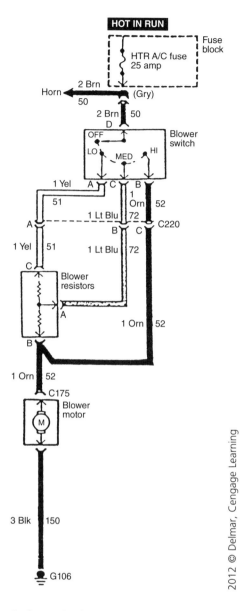

30. Refer to the figure above. The blower motor works in LO and HI positions but does not work on MED. Technician A says the problem could be an open blower resistor. Technician B says the problem could be the blower motor switch. Who is correct?

TASK D.1.2

A. A only

B. B only

C. Both A and B

D. Neither A nor B

Answer A is incorrect. In low speed, current flows through both resistors, so neither one could be open.

Answer B is correct. Only Technician B is correct. The switch provides the path for current to the medium speed resistor. Another possible cause for this problem could be an open in circuit #72, which is the light blue wire leading from the switch to the blower resistor.

Answer C is incorrect. Only Technician B is correct.

Answer D is incorrect. Technician B is correct.

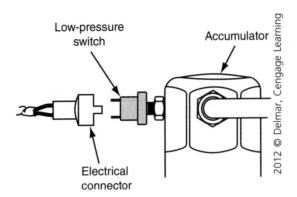

Low-pressure switch

Accumulator

Electrical connector

2012 © Delmar, Cengage Learning

TASK D.1.3

31. All of the following conditions would cause the contacts of the switch in the figure to be "open" EXCEPT:

   A. Cold temperature in the evaporator core.

   B. An inoperative condenser fan.

   C. An empty refrigerant system.

   D. Ambient temperature of 10° F (−12.2° C).

Answer A is incorrect. Cold temperature in the evaporator core would cause the low-side switch contacts to be open, which would keep the evaporator from freezing up.

Answer B is correct. An inoperative condenser fan would cause an increase in high-side pressures. This high pressure would not cause the contacts of the low-side switch to open.

Answer C is incorrect. An empty refrigerant system would cause the low-side switch contacts to open due to the loss of system pressure.

Answer D is incorrect. Ambient temperature of 10° F (−12.2° C) would cause the low-side switch contact to open due to the drop in system pressure. The pressure/temperature relationship causes the pressure to drop as the temperature drops.

TASK D.1.5

32. An electric cooling fan motor can be controlled by any of the following EXCEPT:

   A. An electronic relay.

   B. An independent electronic module.

   C. A multiplexed electronic module.

   D. An air solenoid controller.

Answer A is incorrect. Electric cooling fans are often controlled by an electronic relay.

Answer B is incorrect. A cooling fan module often controls electric cooling fans.

Answer C is incorrect. A multiplexed electronic module often controls electric cooling fans.

Answer D is correct. Electric cooling fan motors are never controlled by an air solenoid controller.

33. Before replacing an electric blend air door actuator, the technician should:

    A.　Ensure that the batteries are removed from the vehicle.

    B.　Ensure that the batteries have a good ground.

    C.　Ensure that the blend door moves freely.

    D.　Ground himself to the vehicle.

TASK D.1.6

Answer A is incorrect. Battery removal has no effect on this service.

Answer B is incorrect. This has no bearing on the removal process.

Answer C is correct. If the blend air door is binding, it could damage the new actuator.

Answer D is incorrect. Blend door actuators do not have sensitive electronics that can be damaged by static electricity.

34. A truck that uses an HVAC system with a vacuum control panel produces no cool air out of the dash outlets. The air only discharges out of the defrost outlets. Which of these items is the most likely cause?

    A.　A leaking dash vacuum switch

    B.　A defective A/C compressor

    C.　Loss of vacuum to the control panel

    D.　A defective heater control valve

TASK D.7

Answer A is incorrect. A leaking dash vacuum switch will hiss and cause some control problems but will not cause total loss of vacuum control to the mode doors.

Answer B is incorrect. A defective A/C compressor would cause total loss of any cool air from any of the vents.

Answer C is correct. The loss of vacuum supply to the control panel results in a fail-safe mode of all air to the defrost outlets.

Answer D is incorrect. A heater control valve failure causes either no cold air or no hot air.

35. All of the following are examples of HVAC control systems types EXCEPT:

    A.　Orifice tube climate control.

    B.　Manual climate control.

    C.　Semi-automatic climate control.

    D.　Automatic climate control.

TASK A.5

Answer A is correct. Many systems use an orifice tube as the metering device, but this is not considered an HVAC control system.

Answer B is incorrect. Manual climate control systems are used on many trucks. These systems require the driver to pick the location and temperature of the desired airflow as well as the blower speed.

Answer C is incorrect. Semi-automatic climate control systems are used on some trucks. These systems have electronic temperature and blower control features but still have a manual mode selector.

Answer D is incorrect. Automatic climate control systems are used on some trucks. These systems have electronic temperature, blower, and mode capabilities. The driver can just choose the desired temperature and the system will activate the necessary systems to achieve the desired temperature.

TASK B.1.5

36. A technician has just recovered the refrigerant from an A/C system. Technician A says that the waste oil bottle should be checked to see how much oil was removed during the recovery process. Technician B says that the recovery weight scale should be checked to see how much refrigerant was removed during the process. Who is correct?

   A.   A only

   B.   B only

   C.   Both A and B

   D.   Neither A nor B

   Answer A is incorrect. Technician B is also correct.

   Answer B is incorrect. Technician A is also correct.

   Answer C is correct. Both Technicians are correct. The waste oil bottle should be checked after recovering an A/C system to monitor how much refrigerant oil was removed during the process. Typically, very little oil is removed during the recovery process. The recovery weight scale should also be checked following a system recovery to see how much refrigerant was removed from the A/C system. This knowledge assists the Technician by showing if the system was charged with the correct amount.

   Answer D is incorrect. Both Technicians are correct.

TASK E.1

37. Technician A says that according to new environmental laws, shut-off valves must be placed in the closed position every time the A/C system is turned off. Technician B says that according to new environmental laws, shut-off valves must be located no more than 12 inches from test hose service ends. Who is correct?

   A.   A only

   B.   B only

   C.   Both A and B

   D.   Neither A nor B

   Answer A is incorrect. Shut-off valves do not have to be closed every time the A/C system is switched off. However, they must be closed when connecting and disconnecting the hoses from the vehicle.

   Answer B is correct. Only Technician B is correct. New environmental laws dictate that shut-off valves must be located no more than 12 inches from test hose service end.

   Answer C is incorrect. Only Technician B is correct.

   Answer D is incorrect. Technician B is correct.

TASK E.2

38. A refrigerant identifier is connected to a truck A/C system and gives the reading of 90 percent R-134a and 10 percent R-12. Technician A says that this system can be safely recovered into the R-134a recovery machine. Technician B says that this system has likely been retrofitted using non-standard procedures. Who is correct?

   A.   A only

   B.   B only

   C.   Both A and B

   D.   Neither A nor B

   Answer A is incorrect. This system has a blend of two different refrigerants which should never be recovered into the main shop A/C equipment.

   Answer B is correct. Only Technician B is correct. This system has R-134a and R-12 which makes it likely that a non-standard retrofit has been performed.

   Answer C is incorrect. Only Technician B is correct.

   Answer D is incorrect. Technician B is correct.

39. Which of these terms correctly describes refrigerant which has been removed from a system and stored in an external container?

TASK E.3

    A. Recycled
    B. Recovered
    C. Reclaimed
    D. Refined

    Answer A is incorrect. Recycled refrigerant has been filtered to remove contaminants and oil.

    Answer B is correct. Refrigerant that has been removed and stored is referred to as recovered. Most new A/C machines recycle the refrigerant as the machine recovers the refrigerant from a vehicle.

    Answer C is incorrect. Reclaimed refrigerant has also been reprocessed to return it to new product quality.

    Answer D is incorrect. Refining is not a process used with refrigerants.

40. The test for non-condensable gases in recovered/recycled refrigerant involves:

TASK E.5

    A. Comparing the pressure of the recovered refrigerant in the container to the theoretical pressure of pure refrigerant at a given temperature.
    B. Comparing the atmospheric pressure to the relative humidity.
    C. Comparing the container pressure with the size of the container.
    D. Testing the refrigerant with a halogen leak detector.

    Answer A is correct. Comparing the pressure of recovered refrigerant to the theoretical pressure of pure refrigerant at a given temperature is the best method of testing for non-condensable gases in refrigerant.

    Answer B is incorrect. Pressure cannot be compared to humidity.

    Answer C is incorrect. Pressure cannot be compared to volume.

    Answer D is incorrect. A halogen leak detector cannot be used to check for non-condensable gases.

PREPARATION EXAM 5—ANSWER KEY

1.	C	21.	A
2.	B	22.	B
3.	C	23.	C
4.	A	24.	B
5.	D	25.	D
6.	B	26.	D
7.	B	27.	B
8.	B	28.	B
9.	D	29.	B
10.	A	30.	B
11.	B	31.	B
12.	D	32.	B
13.	B	33.	D
14.	A	34.	A
15.	C	35.	C
16.	D	36.	A
17.	B	37.	C
18.	B	38.	D
19.	B	39.	D
20.	C	40.	C

PREPARATION EXAM 5—EXPLANATIONS

TASK A.1

1. Which of the following methods would be the LEAST LIKELY way to verify the driver's complaint about an HVAC problem on a truck?

    A. Operate the system in question in the service bay.
    B. Perform a thorough inspection.
    C. Read the repair order.
    D. Perform a road test.

    Answer A is incorrect. Operating the system in the service bay will assist the technician in verifying the driver's complaint.

    Answer B is incorrect. A thorough inspection will assist the technician in verifying the driver's complaint.

    Answer C is correct. Reading the repair order is an important step, but it is not a method of verifying the driver's complaint.

    Answer D is incorrect. Performing a road test will assist the technician in verifying the driver's complaint.

2. An automatic temperature control (ATC) system is being diagnosed. Technician A says that this system must use an electronic control head. Technician B says that this system can be diagnosed with a scan tool. Who is correct?

TASK A.5

   A. A only
   B. B only
   C. Both A and B
   D. Neither A nor B

Answer A is incorrect. Technician B is also correct.

Answer B is incorrect. Technician A is also correct.

Answer C is correct. Both Technicians are correct. ATC systems always use an electronic control head. This control head communicates with the HVAC computer to provide the necessary functionality of ATC systems. Scan tools are often used to diagnose ATC systems. Scan tools can retrieve trouble codes, view live data, and perform output tests on the HVAC system.

Answer D is incorrect. Both Technicians are correct.

3. A compressor has excessive air gap at the clutch drive plate. Technician A says that the clutch could squeal when the compressor engages. Technician B says that the clutch could squeal when the engine is quickly accelerated while running the air conditioning. Who is correct?

TASKS A.2, B.2.4

   A. A only
   B. B only
   C. Both A and B
   D. Neither A nor B

Answer A is incorrect. Technician B is also correct.

Answer B is incorrect. Technician A is also correct.

Answer C is correct. Both Technicians are correct. Excessive air gap at the A/C compressor clutch drive plate could cause the clutch to squeal as the clutch engages, as well as when the engine is quickly accelerated while running the air conditioning. The air gap can be checked with a feeler gauge.

Answer D is incorrect. Both Technicians are correct.

4. Which of the following practices is LEAST LIKELY to be performed during a heater core replacement?

TASK C.11

   A. Inspect the blend and mode doors after reassembling the duct box to assure correct operation.
   B. Completely drain the coolant from the radiator and engine.
   C. Disassemble the instrument panel to gain access to the heater core.
   D. Install hose crimping pliers on the heater hoses.

Answer A is incorrect. The blend and mode doors should be closely inspected after reassembling the duct box to make sure that the doors are in the correct position and move without binding.

Answer B is correct. The engine and radiator do not have to be completely drained to replace the heater core.

Answer C is incorrect. The instrument panel needs to be disassembled to gain access to the heater core. Some trucks require that the complete instrument panel be removed in order to remove the HVAC duct box.

Answer D is incorrect. Installing hose crimping pliers to the heater hoses is a good practice to reduce the loss of coolant during the repair.

TASKS A.4,
B1.1

5. An HVAC system outputs air at a constant temperature, regardless of the temperature setting. What is the most likely cause for this condition?

A. Low refrigerant charge

B. Low coolant level

C. Compressor clutch failure

D. Broken blend door cable

Answer A is incorrect. A low refrigerant charge will not affect heater temperature control.

Answer B is incorrect. Even with a very low coolant level, cooling will take place with the A/C on.

Answer C is incorrect. A compressor clutch failure will not affect heater temperature control.

Answer D is correct. A broken blend door cable will prevent the control of the HVAC system output air. The blend door moves to route the air through or around the heater core. When the temperature lever is set to hot, the blend door routes all of the air past the heater core. When the temperature lever is in the full cold position, the blend door blocks any air from passing by the heater core.

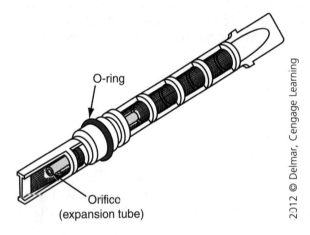

O-ring

Orifice
(expansion tube)

2012 © Delmar, Cengage Learning

TASK A.4

6. Referring to the figure, Technician A says that the device is an orifice tube and is used on truck systems that use a receiver/drier. Technician B says that the device meters refrigerant into the evaporator core. Who is correct?

A. A only

B. B only

C. Both A and B

D. Neither A nor B

Answer A is incorrect. Orifice tube systems use an accumulator/drier.

Answer B is correct. Only Technician B is correct. The orifice tube is a metering device that meters high-pressure liquid refrigerant into the evaporator core. After passing into the evaporator core, the refrigerant begins to boil into a vapor.

Answer C is incorrect. Only Technician B is correct.

Answer D is incorrect. Technician B is correct.

7. Which of these characteristics does the R-134a refrigerant possess?

TASK B.1.2

A. No odor

B. Faint ether-like odor

C. Strong rotten egg odor

D. Cabbage-like odor

Answer A is incorrect. R134a does have an ether-like odor.

Answer B is correct. R-134a has a faint ether-like odor. It is possible for a customer to comment that they have an ether-like smell in their vehicle if the evaporator core has a large leak.

Answer C is incorrect. R-134a does not have a strong rotten egg odor.

Answer D is incorrect. R-134a does not have a cabbage-like odor.

8. Which of the following procedures would be LEAST LIKELY used to determine the freeze protection level of the engine coolant?

TASK C.3

A. Test strips

B. Visual inspection

C. Hydrometer

D. Refractometer

Answer A is incorrect. Test strips can be used to test the freeze protection as well as for the presence of certain chemicals in the engine coolant.

Answer B is correct. A visual inspection of the engine coolant would not provide the freeze protection level. A test instrument or test strips is required to measure freeze protection.

Answer C is incorrect. A hydrometer can be used to test the freeze protection of the engine coolant. A small sample is drawn into the hydrometer to cause an indicator to reveal the freeze protection point.

Answer D is incorrect. A refractometer can be used to test the freeze protection of the engine coolant. A small drop of the coolant is inserted onto the test window of the refractometer. This is the most accurate coolant freeze protection method.

9. An A/C system has been open to the atmosphere during a repair. What is the minimum recommended length of time that the vacuum pump should be operated for evacuation?

TASK B.1.6

A. 20 minutes

B. 10 minutes

C. 15 minutes

D. 30 minutes

Answer A is incorrect. Twenty minutes is not enough time to evacuate.

Answer B is incorrect. Ten minutes is not enough time to evacuate.

Answer C is incorrect. Fifteen minutes is not enough time to evacuate.

Answer D is correct. To ensure that the entire system is under deep enough vacuum to remove all moisture, the pump should be run for at least 30 minutes. After evacuating for 30 minutes, the technician should let the system sit with the vacuum pump turned off and make sure that the vacuum is maintained. If the system does hold a vacuum, there are no major leaks in the system and it can be charged with refrigerant.

TASK B.1.8

10. Technician A says the A/C system can be charged through the low side with the system running. Technician B says inverting the refrigerant container causes low-pressure refrigerant vapor to be charged into the system. Who is correct?

    A. A only

    B. B only

    C. Both A and B

    D. Neither A nor B

Answer A is correct. Only Technician A is correct. It is acceptable to charge an A/C system into the low side while the system is running. Caution must be taken to make sure that the high-side manifold valves are closed during this process to make sure that high pressure does not enter the charging tank.

Answer B is incorrect. Inverting the refrigerant container will cause low-pressure liquid to be charged into the system.

Answer C is incorrect. Only Technician A is correct.

Answer D is incorrect. Technician A is correct.

TASK D.3

11. All of the following electronic devices could be used to control the operation on truck A/C systems EXCEPT:

    A. Pressure cycling switch.

    B. Engine oil temperature sensor.

    C. Evaporator temperature switch.

    D. Binary pressure switch.

Answer A is incorrect. The pressure cycling switch is located in the low side of the refrigerant system. This switch opens when the pressure in the low side of the system drops below approximately 25 psi.

Answer B is correct. The engine oil temperature sensor senses the temperature of the engine oil and is not used to control the operation of the truck A/C system.

Answer C is incorrect. The evaporator temperature switch is located near the evaporator core. This switch opens when the temperature of the evaporator core drops to near freezing temperature.

Answer D is incorrect. The binary pressure switch is located in the high side of the refrigerant system. This switch opens when the pressure in the high side rises above a preset level or drops below a preset level.

12. Technician A says that some HVAC electrical control panels have replaceable bulbs. Technician B says that the HVAC electrical control panel can be replaced without removing the instrument panel trim bezel. Who is correct?

TASK D.7

A. A only

B. B only

C. Both A and B

D. Neither A nor B

Answer A is correct. Only Technician A is correct. The backlight bulbs can be serviced on some HVAC electrical control panels. The technician should replace all of the backlight bulbs when completing this service.

Answer B is incorrect. The instrument panel trim bezel will normally have to be removed during the replacement of the HVAC electrical control panel.

Answer C is incorrect. Only Technician A is correct.

Answer D is incorrect. Technician A is correct.

13. The A/C compressor clutch will not engage in any mode. The clutch engages when a technician installs a jumper wire across the terminals of the low-pressure cut-out switch connector. Technician A says that the low-pressure cut-out switch must be defective. Technician B says that the refrigerant charge could be low. Who is correct?

TASK B.2.1

A. A only

B. B only

C. Both A and B

D. Neither A nor B

Answer A is incorrect. The low-pressure cut-out switch is not necessarily at fault. The system could be low on refrigerant.

Answer B is correct. Only Technician B is correct. A low refrigerant charge will cause this effect. The low-pressure cut-out switch opens its contacts when the pressure drops below 20–25 psi. The switch acts as a de-icing device when the system is operating. The switch also serves as a low refrigerant protection device that keeps the compressor from turning on when the refrigerant gets low.

Answer C is incorrect. Only Technician B is correct.

Answer D is incorrect. Technician B is correct.

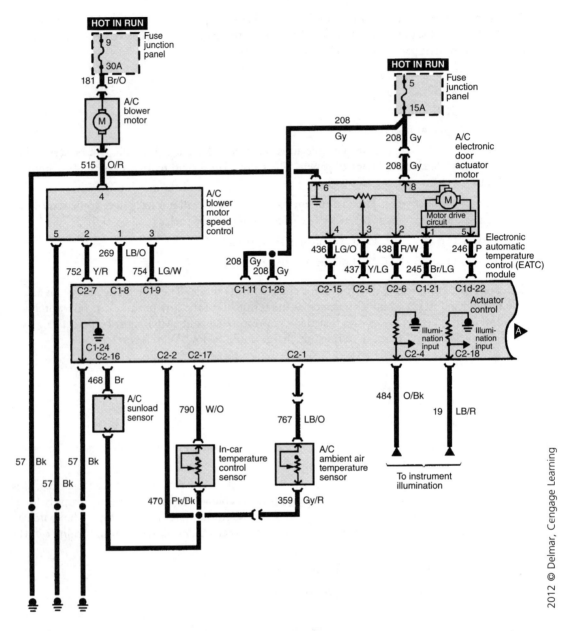

TASK D.1

14. Referring to the figure, when measuring the voltage drop in the A/C computer ground as shown, the technician connects a voltmeter from computer terminal C1-24 internal ground to the external ground. With the ignition on, the maximum allowable voltage drop should be which of the following?

  A. 0.05 volts
  B. 0.3 volts
  C. 0.5 volts
  D. 0.8 volts

  Answer A is correct. A voltage drop of 0.05 volts is the lowest of the choices and represents the maximum voltage drop allowed across a low-amperage ground circuit.

  Answer B is incorrect. The drop is too large, indicating high resistance in the ground circuit.

  Answer C is incorrect. The drop is too large, indicating high resistance in the ground circuit. This number represents the maximum voltage drop in a battery cable with the engine cranking.

  Answer D is incorrect. The drop is too large, indicating high resistance in the ground circuit.

15. Technician A says that you can test the A/C compressor clutch coil with an ohmmeter. Technician B says that connecting battery power to one terminal of the coil and grounding the other terminal can test the A/C compressor clutch coil. Who is correct?

TASK B.2.4

    A. A only

    B. B only

    C. Both A and B

    D. Neither A nor B

Answer A is incorrect. Technician B is also correct.

Answer B is incorrect. Technician A is also correct.

Answer C is correct. Both Technicians are correct. The resistance of the clutch coil can be tested with an ohmmeter. Another test that can be performed is supplying power and ground to the coil to see if the clutch will engage.

Answer D is incorrect. Both Technicians are correct.

16. Which of the following statements is true about checking the A/C compressor lubricant level?

    A. The A/C system must first be evacuated for at least 30 minutes.

    B. The compressor must not be operated for at least 24 hours before checking the lubricant level.

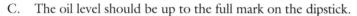

TASK B.2.5

    C. The oil level should be up to the full mark on the dipstick.

    D. The old lubricant must be measured before the new lubricant is added.

Answer A is incorrect. The A/C system does not need to be evacuated for 30 minutes prior to checking the lubricant. The refrigerant does need to be recovered with an approved recovery machine.

Answer B is incorrect. The compressor needs to be operated just prior to recovering the refrigerant to ensure that the lubricant is properly distributed through the system.

Answer C is incorrect. Modern compressors do not have a dipstick to allow the oil to be checked.

Answer D is correct. The old lubricant should be measured before new lubricant is added. After adding the new lubricant, the compressor should be turned several revolutions by hand to help avoid compressor damage when it is first turned on.

17. Technician A says that the compressor reed valves can often be replaced without discharging the A/C system. Technician B says that damaged compressor reed valves can send debris into the condenser. Who is correct?

TASKS B.2.6,
B.3.4

    A. A only

    B. B only

    C. Both A and B

    D. Neither A nor B

Answer A is incorrect. The compressor will need to be replaced if the reed valves become damaged. The condenser will also need to be replaced due to the debris from the damaged reed valves being pushed into the condenser.

Answer B is correct. Only Technician B is correct. Damaged compressor reed valves usually end up in the condenser. The compressor and the condenser will need to be replaced to correct this problem.

Answer C is incorrect. Only Technician B is correct.

Answer D is incorrect. Technician B is correct.

TASK D.6

18. All of the following statements about a computer-controlled A/C system are correct EXCEPT:

   A. Some actuator motors are calibrated automatically in the self-diagnostic mode.

   B. A/C diagnostic trouble codes (DTC) represent the exact fault in a specific component.

   C. The actuator control rods must be calibrated manually on some systems.

   D. The actuator motor control rods should only require adjustment after motor replacement or adjustment.

Answer A is incorrect. Some actuator motors can be calibrated by putting the control head in self-diagnostic mode.

Answer B is correct. A/C diagnostic trouble codes (DTC) represent a fault in a specific system, not a component.

Answer C is incorrect. The control rods do have to be manually adjusted on a few A/C systems.

Answer D is incorrect. The only time a technician has to adjust the motor control rods is after motor replacement or adjustment.

TASK B.3.3

19. Airflow through the A/C condenser is significantly affected by all of the following EXCEPT:

   A. Debris trapped in the fins.

   B. Relative humidity of the outside air.

   C. Bent cooling fins.

   D. Vehicle speed.

Answer A is incorrect. Debris trapped in the condenser fins will significantly affect airflow through the condenser.

Answer B is correct. Relative humidity will not significantly affect airflow through the condenser. However, the humidity level greatly affects the pressures in the high side of the system. High humidity levels cause the pressure in the high side of the system to elevate dramatically.

Answer C is incorrect. Bent fins will affect airflow through the condenser.

Answer D is incorrect. Vehicle speed will affect airflow through the condenser.

TASK C.10

20. Technician A says that the heater control valve can be controlled by oil pressure. Technician B says that the heater control valve can be cable operated. Who is correct?

   A. A only

   B. B only

   C. Both A and B

   D. Neither A nor B

Answer A is incorrect. The heater control valves are not controlled by oil pressure.

Answer B is correct. Only Technician B is correct. The heater control valves may be cable operated. Other methods for operating heater control valves are vacuum, air pressure, and electrical solenoids.

Answer C is incorrect. Only Technician B is correct.

Answer D is incorrect. Technician B is correct.

21. A screen is located in the orifice tube of an A/C system. Technician A says that the screen is a filter used to prevent particulate from circulating through the system. Technician B says that the screen is used to improve atomization of the refrigerant. Who is correct?

TASK B.3.7

A. A only

B. B only

C. Both A and B

D. Neither A nor B

Answer A is correct. Only Technician A is correct. The orifice tube screen is installed to prevent particulate from circulating through the system in the event that the desiccant bag in the accumulator breaks down.

Answer B is incorrect. Atomization is not important to the evaporation of refrigerant.

Answer C is incorrect. Only Technician A is correct.

Answer D is incorrect. Technician A is correct.

22. The inside of a truck windshield has an oily film and the A/C cooling is poor. Technician A says a plugged HVAC evaporator drain may cause this oil film. Technician B says this film may be caused by a large leak in the evaporator core. Who is correct?

TASKS A.3,
B.3.8

A. A only

B. B only

C. Both A and B

D. Neither A nor B

Answer A is incorrect. A plugged evaporator drain may cause windshield fogging, but not an oily film on the windshield.

Answer B is correct. Only Technician B is correct. A refrigerant leak in the evaporator core may allow some refrigerant and oil to leak, thus causing a thin oily film on the windshield.

Answer C is incorrect. Only Technician B is correct.

Answer D is incorrect. Technician B is correct.

23. A truck is being diagnosed for an A/C system problem. The operating pressures are checked. The results show the low-side pressure is too high and the high-side pressure is too low. Technician A says that the compressor may have a faulty reed valve. Technician B says that an overcharge of refrigerant oil is a possible cause. Who is correct?

TASK B.1.3

A. A only

B. B only

C. Both A and B

D. Neither A nor B

Answer A is correct. Only Technician A is correct. A faulty reed valve could cause the high-side pressure to be low and the low-side pressure to be high.

Answer B is incorrect. An overcharge of refrigerant oil will not cause these symptoms, but will reduce the cooling capabilities of the system. The likely pressures that would be present on a system with too much oil are high pressure readings in both sides of the system.

Answer C is incorrect. Only Technician A is correct.

Answer D is incorrect. Technician A is correct.

TASK C.5

24. Which of the following is the expected result from raising the pressure in the cooling system?

    A. Lower coolant boiling point

    B. Elevated coolant boiling point

    C. Corrosion prevention in the cooling system

    D. No change in the coolant boiling point

Answer A is incorrect. Raising the pressure of the cooling system raises the boiling point of the coolant.

Answer B is correct. Raising the pressure compresses the rate of gas expansion, thereby increasing the temperature to reach the boiling point. Each 1 psi of pressure added to the cooling system results in the boiling point of the coolant being raised by 3° F.

Answer C is incorrect. Raising the pressure in the cooling system does not prevent corrosion. Antifreeze contains corrosion inhibiters that help prevent corrosion.

Answer D is incorrect. Raising the pressure raises the boiling point 3° F for each psi of pressure that is added.

TASKS C.5, C.7

25. All of the following are good methods of verifying that an engine thermostat opens EXCEPT:

    A. Feeling the upper radiator hose.

    B. Watching the temperature gauge.

    C. Watching for motion in the upper radiator tank.

    D. Watching the surge tank.

Answer A is incorrect. When the thermostat opens, the upper radiator hose rapidly gets warm.

Answer B is incorrect. When the thermostat opens, the temperature gauge rises until it indicates normal operating temperature.

Answer C is incorrect. When the thermostat is open, there is obvious circulation motion in the upper radiator tank.

Answer D is correct. Thermostat opening often is not noticeable in the surge tank.

TASK C.9

26. A cooling fan can be controlled by any of the following EXCEPT:

    A. An air clutch.

    B. A thermostatic spring and fan clutch fluid.

    C. An electrically actuated clutch.

    D. A hydraulic switch.

Answer A is incorrect. Many fan clutches are air operated.

Answer B is incorrect. Some fan clutches are operated by a thermostatic spring and fan clutch fluid.

Answer C is incorrect. Some fans use an electrically actuated clutch.

Answer D is correct. Hydraulic switches are rarely used for cooling fans.

27. When an expansion valve is properly installed, where should the thermal bulb and capillary tube be positioned?

TASK B.3.6

    A. Fastened to the condenser fins using epoxy

    B. Located in the accumulator

    C. Held in contact with the evaporator outlet using insulating tape

    D. As an integral part of the orifice tube assembly

Answer A is incorrect. The expansion valve is located near the evaporator, not the condenser.

Answer B is incorrect. An accumulator is used with a CCOT system, which does not use a thermal bulb or capillary tube.

Answer C is correct. The thermal bulb and capillary tube are part of the thermal expansion valve located near the evaporator inside of the duct housing. The thermal bulb and capillary tube sense the temperature of the outlet line of the evaporator core. When the thermal bulb senses warm temperatures, the TXV opens to allow more refrigerant into the evaporator core. When the thermal bulb senses cold temperatures, the TXV closes to restrict refrigerant flow into the evaporator.

Answer D is incorrect. A CCOT system does not use a thermal bulb or capillary tube.

28. What could a faint ether-like odor coming from the panel vents in the NORMAL A/C mode indicate?

TASK A.3

    A. The evaporator core is leaking R-134a.

    B. The evaporator core is leaking R-12.

    C. The cold starting system is malfunctioning.

    D. The heater core is leaking.

Answer A is correct. R-134a has a faint ether-like odor. However, it is very rare to actually be able to smell a refrigerant leak. A leak of this size would drain the system in a short time.

Answer B is incorrect. R-12 is an odorless gas that would not be detectable to a person.

Answer C is incorrect. HVAC system input air is not drawn from under the hood or the cab.

Answer D is incorrect. A leaking heater core will produce a sweet odor, not an ether-like odor.

29. A truck is being diagnosed for having three compressor failures in a six-month period. Each compressor has failed in the area of the front drive clutch overheating. Technician A says that a weak battery pack could be causing this problem. Technician B says that a faulty A/C clutch relay could be causing the problem. Who is correct?

TASK D.1

    A. A only

    B. B only

    C. Both A and B

    D. Neither A nor B

Answer A is incorrect. A weak battery pack would not likely cause a repeated compressor failure. There would be other starting-related problems associated with weak batteries.

Answer B is correct. Only Technician B is correct. An A/C clutch relay with burnt contacts could be at fault. If the A/C compressor clutch coil does not receive system voltage, then the magnetic field will be weak and cause the front clutch plate to slip.

Answer C is incorrect. Only Technician B is correct.

Answer D is incorrect. Technician B is correct.

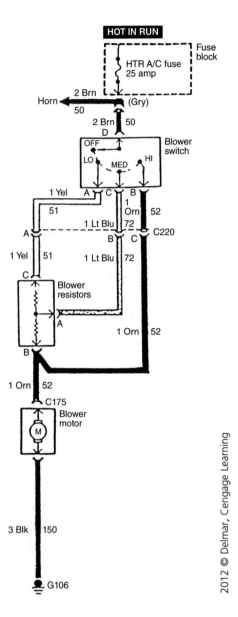

TASK D.2

30. Referring to the figure above, Technician A says that the blower motor has four speeds. Technician B says that the blower resistor has two resistors. Who is correct?

A.   A only

B.   B only

C.   Both A and B

D.   Neither A nor B

Answer A is incorrect. The blower motor does not have four speeds. The blower switch has low speed, medium speed, and high speed.

Answer B is correct. Only Technician B is correct. The blower resistor has two resistors. Power is routed through both resistors when the blower switch is turned to low speed. Power is routed through one of the resistors when the blower switch is turned to medium speed.

Answer C is incorrect. Only Technician B is correct.

Answer D is incorrect. Technician B is correct.

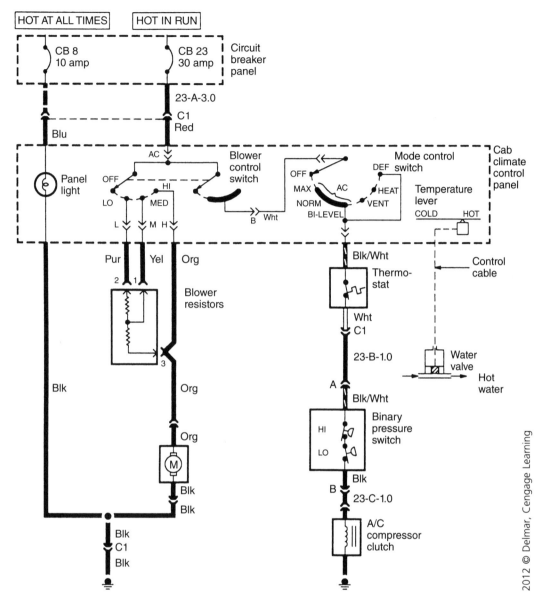

2012 © Delmar, Cengage Learning

31. Referring to the figure above, the compressor clutch is inoperative in the DEF mode but operates properly in all other A/C modes. Technician A says the low side of the binary switch may have an open circuit. Technician B says the defrost switch contacts may have an open circuit. Who is correct?

TASK D.3

　A.　A only

　B.　B only

　C.　Both A and B

　D.　Neither A nor B

Answer A is incorrect. If the binary switch were open then the system would not operate in the other A/C modes.

Answer B is correct. Only Technician B is correct. The defrost switch contacts could be the cause of the compressor clutch not operating in defrost mode.

Answer C is incorrect. Only Technician B is correct.

Answer D is incorrect. Technician B is correct.

TASK B.3.1

32. Technician A says that the evaporator must be removed from the vehicle if the presence of evaporator lubricant is to be checked. Technician B says that the refrigeration oil is distributed throughout the A/C system. Who is correct?

    A. A only

    B. B only

    C. Both A and B

    D. Neither A nor B

Answer A is incorrect. The oil level could be verified by flushing the complete system and then adding the specified amount of oil to each component.

Answer B is correct. Only Technician B is correct. The refrigeration oil is distributed throughout the system.

Answer C is incorrect. Only Technician B is correct.

Answer D is incorrect. Technician B is correct.

TASK D.3

33. Which of the following electronic devices would be LEAST LIKELY to control the operation of a truck A/C system?

    A. Low-pressure switch

    B. Dual pressure switch

    C. A/C pressure sensor

    D. Fuel temperature sensor

Answer A is incorrect. The low-pressure switch is located in the low side of the refrigerant system. This switch opens when the pressure in the low side of the system drops below approximately 25 psi.

Answer B is incorrect. The dual pressure switch is located in the high side of the refrigerant system. This switch opens when the pressure in the high side rises above a preset level or drops below a preset level.

Answer C is incorrect. The A/C pressure sensor senses the refrigerant pressure in the high side of the A/C system.

Answer D is correct. The fuel temperature sensor senses fuel temperature and does not control the operation of the truck A/C system.

TASK D.7

34. Which of the following procedures would be the most likely method of repairing an air leak in the HVAC control panel?

    A. Replacement of the control panel

    B. Replacement of the pintle o-rings

    C. Repacking the selector body with grease

    D. Replacing the selector levers

Answer A is correct. Replacement is the only good repair for an HVAC control panel with an air leak.

Answer B is incorrect. This process is not advisable because the head is not designed to be serviced in this way.

Answer C is incorrect. Applying grease will not improve the operation of the leaking head.

Answer D is incorrect. The selector levers have nothing to do with a leaking control panel.

35. A truck that has an automatic temperature control (ATC) system is being diagnosed. Technician A says that these systems have a self-diagnostic system that sets trouble codes when certain problems occur. Technician B says that ATC trouble codes can be retrieved with a scan tool. Who is correct?

TASKS A.5, D.11

    A. A only

    B. B only

    C. Both A and B

    D. Neither A nor B

Answer A is incorrect. Technician B is also correct.

Answer B is incorrect. Technician A is also correct.

Answer C is correct. Both Technicians are correct. ATC systems are very advanced in design and operation. These systems typically have a self-diagnostic system that sets trouble codes when certain problems occur. Scan tools can be used to retrieve trouble codes from these systems. In addition, scan tools can be used to view live data and perform output tests on the systems.

Answer D is incorrect. Both Technicians are correct.

36. A heavy truck with a faulty A/C pressure sensor is being diagnosed. Technician A says that this device is a three-wire feedback sensor that is used to signal A/C system pressure to a control module. Technician B says that this device is typically mounted in the low side of the A/C system. Who is correct?

TASK B.2.2

    A. A only

    B. B only

    C. Both A and B

    D. Neither A nor B

Answer A is correct. Only Technician A is correct. A/C pressure sensors are three-wire devices used on the high side of A/C systems to signal A/C system pressure to a control module. This varying voltage is a more accurate method of controlling the A/C system. The range of the voltage is approximately 1 to 3 volts during operation.

Answer B is incorrect. A/C pressure sensors are typically located on the high side of the refrigerant system.

Answer C is incorrect. Only Technician A is correct.

Answer D is incorrect. Technician A is correct.

37. Technician A says that all A/C repair shops are required to use SAE-approved recovery equipment. Technician B says that all individuals who service A/C systems must be certified by a recognized body on how to properly handle refrigerants. Who is correct?

TASK B.1.12

    A. A only

    B. B only

    C. Both A and B

    D. Neither A nor B

Answer A is incorrect. Technician B is also correct.

Answer B is incorrect. Technician A is also correct.

Answer C is correct. Both Technicians are correct. All A/C repair shops do have to use SAE-approved A/C equipment. The service technicians must also pass certification tests distributed by recognized bodies showing that they can handle A/C refrigerants properly.

Answer D is incorrect. Both Technicians are correct.

TASK B.1.10

38. Federal laws require safe refrigerant handling procedures. Technician A says that stored refrigerant should be kept warm by providing a heat source near the storage area. Technician B says that recovered refrigerant should be kept in a DOT 39 cylinder. Who is correct?

    A. A only
    B. B only
    C. Both A and B
    D. Neither A nor B

    Answer A is incorrect. Stored refrigerant should never be stored near a heat source due to its pressure/temperature characteristic. The pressure increases as the temperature rises.

    Answer B is incorrect. Recovered refrigerant should never be kept in a DOT 39 cylinder. Always use a DOT 4BW cylinder for storing recovered refrigerant.

    Answer C is incorrect. Neither Technician is correct.

    Answer D is correct. Neither Technician is correct. It is dangerous to store refrigerant near a heat source. Always use a DOT 4BW cylinder to store recovered refrigerant.

TASK D.12

39. All of the following steps should be performed after an HVAC repair has been made EXCEPT:

    A. A thorough visual inspection.
    B. Clearing diagnostic codes.
    C. Operating the system to check performance.
    D. Recovering the refrigerant.

    Answer A is incorrect. It is a good practice to perform a thorough visual inspection after each HVAC repair. The technician should inspect all of the components, fasteners, connections, and wires to make sure that all items are in place and ready to perform at a high level.

    Answer B is incorrect. The technician should always clear all diagnostic codes after each repair in order to assure that the truck leaves the shop with a clear computer.

    Answer C is incorrect. The technician should operate the system to make sure it is performing up to specifications.

    Answer D is correct. The system does not need to be recovered after an HVAC repair has been made. Recovery is necessary when the A/C system needs to have the refrigerant removed to repair or replace a refrigerant system component.

TASKS A.2, B.2.4, C.6

40. Technician A says that a failed water pump bearing produces a metallic knock at the rear of the engine. Technician B says that a failed compressor pulley bearing will produce noise when the compressor is not engaged. Who is correct?

    A. A only
    B. B only
    C. Both A and B
    D. Neither A nor B

    Answer A is incorrect. A failed water pump bearing would not produce a knock at the rear of the engine. A water pump would produce noise at the front of the engine.

    Answer B is correct. Only Technician B is correct. A failed compressor pulley bearing would produce noise when the compressor is not engaged due to one section of the bearing being stationary and the other section rotating.

    Answer C is incorrect. Only Technician B is correct.

    Answer D is incorrect. Technician B is correct.

PREPARATION EXAM 6—ANSWER KEY

1.	D	21.	A
2.	A	22.	C
3.	B	23.	A
4.	D	24.	C
5.	A	25.	A
6.	B	26.	D
7.	A	27.	A
8.	D	28.	B
9.	A	29.	B
10.	B	30.	A
11.	B	31.	C
12.	C	32.	B
13.	D	33.	C
14.	A	34.	A
15.	A	35.	A
16.	C	36.	D
17.	C	37.	C
18.	A	38.	C
19.	D	39.	B
20.	B	40.	B

PREPARATION EXAM 6—EXPLANATIONS

TASK A.1

1. Which of the following examples would be a benefit of reviewing the past maintenance documents for a truck?

    A. Determining how much money the owner has spent on service
    B. Finding out how to perform the road test
    C. Assisting in performing the visual inspection
    D. Determining if the truck has had past HVAC-related repairs

    Answer A is incorrect. Discovering the service budget for a truck is not a benefit from inspecting the past maintenance documents.

    Answer B is incorrect. Reviewing the past maintenance documents would not assist the technician in performing the road test.

    Answer C is incorrect. Reviewing the past maintenance documents would not assist the technician in performing a visual inspection.

    Answer D is correct. Reviewing the past maintenance documents would assist the technician by revealing the past HVAC-related repairs on the truck.

TASKS A.5, D.11

2. A truck that has an automatic temperature control (ATC) system is being diagnosed. Technician A says that ATC trouble codes are retrieved by depressing a sequence of buttons on the ATC control head. Technician B says that the trouble code reveals exactly what needs to be repaired in an ATC system. Who is correct?

    A. A only

    B. B only

    C. Both A and B

    D. Neither A nor B

    Answer A is correct. Only Technician A is correct. Many ATC systems are designed to display ATC trouble codes when a sequence of buttons on the ATC control head is depressed. The trouble codes are then investigated by performing extensive tests on the system to find a root cause of the problem.

    Answer B is incorrect. Trouble codes do not reveal exactly what needs to be repaired in an ATC system. These codes give the technician an area to begin the troubleshooting process. The technician will usually have to use a database to find a troubleshooting process for each code. The troubleshooting process leads the technician in a logical sequence to the root cause of the problem.

    Answer C is incorrect. Only Technician A is correct.

    Answer D is incorrect. Technician A is correct.

TASK A.3

3. Technician A says that the line exiting the condenser should be hotter than the line entering the condenser. Technician B says that the suction line should be cold to the touch when the A/C system is operating. Who is correct?

    A. A only

    B. B only

    C. Both A and B

    D. Neither A nor B

    Answer A is incorrect. The line exiting the condenser should be from 20° to 50° F cooler than the inlet line. The condenser is a heat exchanger that causes the refrigerant to release heat, which causes the vapor to condense into a liquid.

    Answer B is correct. Only Technician B is correct. The suction line is typically cold during normal A/C operation since it is a low-pressure vapor.

    Answer C is incorrect. Only Technician B is correct.

    Answer D is incorrect. Technician B is correct.

4. A truck technician discovers a heater core leak that requires its replacement. Technician A says that the entire cooling system will need to be drained prior to removal of the heater core. Technician B says that the replacement heater core should be installed without insulation tape. Who is correct?

    A. A only
    B. B only
    C. Both A and B
    D. Neither A nor B

TASK C.11

Answer A is incorrect. The cooling system does not have to be completely drained to replace the heater core.

Answer B is incorrect. The replacement heater core will need to be properly insulated during installation in order to prevent the duct air from by-passing the core.

Answer C is incorrect. Neither Technician is correct.

Answer D is correct. Neither Technician is correct. The cooling system does not have to be completely drained to replace the heater core. Hose crimping pliers can be used to block off the heater hoses to prevent substantial loss of coolant during this repair. It is important to carefully insulate the replacement heater core to assure that the duct air will not by-pass the core.

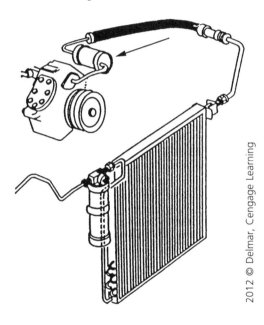

2012 © Delmar, Cengage Learning

5. Referring to the figure above, Technician A says the device indicated by the arrow is used on some trucks to reduce compressor noise. Technician B says that the device indicated by the arrow can be flushed out if it gets restricted. Who is correct?

    A. A only
    B. B only
    C. Both A and B
    D. Neither A nor B

TASK A.3

Answer A is correct. Only Technician A is correct. The muffler in the figure is used by some manufacturers to reduce compressor noise.

Answer B is incorrect. The muffler in the figure is not a component that can be effectively flushed if it becomes restricted. The discharge line will have to be replaced if the muffler becomes restricted.

Answer C is incorrect. Only Technician A is correct.

Answer D is incorrect. Technician A is correct.

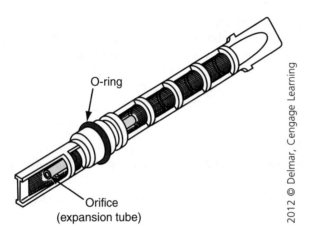

O-ring

Orifice
(expansion tube)

2012 © Delmar, Cengage Learning

TASK A.4

6. What is the most likely drying component to be used with the metering device in the figure above?

    A. Valves in receiver/drier

    B. Accumulator/drier

    C. Muffler drier

    D. Receiver/drier

    Answer A is incorrect. This type of drier was used on early model A/C systems that used suction throttling technology.

    Answer B is correct. Orifice tube A/C systems use an accumulator/drier that is located between the evaporator core and the compressor.

    Answer C is incorrect. A muffler does not typically contain a drier.

    Answer D is incorrect. Thermal expansion valve A/C systems use a receiver/drier that is located in the liquid line of the A/C system.

TASK B.1.2

7. Technician A says that the easiest way to identify the type of refrigerant that a system should use is to observe the service port fittings. Technician B says that R-134a is the only refrigerant that may be vented to the atmosphere. Who is correct?

    A. A only

    B. B only

    C. Both A and B

    D. Neither A nor B

    Answer A is correct. Only Technician A is correct. The service port fittings are different and will easily identify which refrigerant should be used.

    Answer B is incorrect. R-134a should never be vented to the atmosphere.

    Answer C is incorrect. Only Technician A is correct.

    Answer D is incorrect. Technician A is correct.

8. When a technician pressure tests a cooling system, there are no obvious external leaks but the system cannot maintain pressure. Which of the following is the most likely cause of this problem?

    A. Leaking evaporator

    B. Defective heater valve

    C. Stuck open thermostat

    D. Blown head gasket in the engine

TASK C.3

    Answer A is incorrect. The evaporator is not a cooling system component.

    Answer B is incorrect. A defective heater valve would have been found during the external inspection.

    Answer C is incorrect. A stuck-open thermostat will cause excessive cooling of the engine.

    Answer D is correct. A blown head gasket is the most likely cause of an internal engine coolant leak.

9. Which of the following items is most likely to cause elevated high-side pressure in an A/C system?

    A. Restricted airflow through condenser

    B. Stuck open thermostat

    C. Leaking thermal bulb

    D. Inoperative compressor

TASK B.1.3

    Answer A is correct. Restricted airflow through the condenser will cause elevated high-side pressure.

    Answer B is incorrect. A thermostat that is stuck open will cause poor heater performance.

    Answer C is incorrect. The thermal bulb will not leak.

    Answer D is incorrect. An inoperative compressor would cause the pressures to be equal on both sides of the A/C system.

10. Where would you position the leak detector sensor to detect a refrigerant leak?

    A. Within three inches of the fitting

    B. Just below the fitting

    C. Right next to the fitting

    D. Just above the fitting

TASK B.1.4

    Answer A is incorrect. The sensor tip should be as close as possible to the fitting.

    Answer B is correct. Refrigerant is heavier than air, so it is most easily detected just below a leaking fitting.

    Answer C is incorrect. The sensor probe should be just below the fitting.

    Answer D is incorrect. The leak could go undetected using this method. The sensor probe should be just below the fitting.

TASK B.1.6

11. Technician A says that moisture can be removed from the A/C system after charging the system with new refrigerant. Technician B says moisture that enters the A/C circuit will be harmful to the system and cause poor performance. Who is correct?

    A. A only
    B. B only
    C. Both A and B
    D. Neither A nor B

    Answer A is incorrect. Moisture cannot be removed from the A/C system after charging the system with new refrigerant. All systems have a drying device, however, that should capture any trace amounts of moisture that enters the system.

    Answer B is correct. Only Technician B is correct. Moisture that enters the A/C system will be harmful to the system and cause poor performance because of internal erosion and the possibility of the moisture turning to ice near the expansion device.

    Answer C is incorrect. Only Technician B is correct.

    Answer D is incorrect. Technician B is correct.

12. Which of the following functions would be LEAST LIKELY to be performed with a scan tool on a climate control computer?

    A. Display sensor data.
    B. Calibrate the air handling door actuators.
    C. Reprogram the blower motor.
    D. Display switch data.

    Answer A is incorrect. A scan tool can be used to display sensor data for the technician. The sensor data will often reveal problems in the climate control system.

    Answer B is incorrect. A scan tool can be used to calibrate the air handling doors of the climate control system. These actuators have to be calibrated after being replaced.

    Answer C is correct. The blower motor does not need to be reprogrammed since it is simply a permanent magnet device that has no logic functions.

    Answer D is incorrect. A scan tool can be used to display switch data for the technician. The switch data will often reveal problems in the climate control system.

TASK D.11

TASK B.2.2

13. Which of the following components would be LEAST LIKELY used as a de-icing device in the A/C system?

    A. Variable displacement compressor
    B. Pressure cycling switch
    C. Evaporator temperature sensor
    D. Pressure release valve

    Answer A is incorrect. A variable displacement compressor is used to regulate pressure and temperature in the A/C system on some trucks. The system will not freeze because this type of compressor curbs the output when low-side pressure gets low.

    Answer B is incorrect. A pressure cycling switch is used as a de-icing device on some trucks. This device opens to cause the compressor to turn off when low-side pressure drops to approximately 25 psi (172 kPa).

    Answer C is incorrect. An evaporator temperature sensor is used as a de-icing device on some trucks. This thermistor monitors the temperature in the evaporator and causes the compressor to turn off if the temperature drops to near 32° F (0° C).

    Answer D is correct. A pressure release valve would not be used as a de-icing device on an A/C system. This device is a mechanical valve that releases pressure if the high-side pressure rises above a preset limit.

14. A serpentine belt is being replaced on a diesel-powered truck and the tensioner will not snap back after being released. Technician A says that the tensioner needs to be greased to get it to function correctly. Technician B says that the idler pulley is jammed and will need to be replaced. Who is correct?

TASK B.2.3

   A. A only
   B. B only
   C. Both A and B
   D. Neither A nor B

   Answer A is incorrect. It is not a normal practice to grease the tensioner assembly. The tensioner will need to be replaced.

   Answer B is incorrect. The idler pulley is a separate device from the tensioner pulley.

   Answer C is incorrect. Neither Technician is correct.

   Answer D is correct. Neither Technician is correct. A locked tensioner should be replaced. There is no evidence that the idler pulley is defective.

15. Which of the following A/C gauge set readings would most likely indicate the A/C compressor reed valves are worn out?

   A. High-side pressure too low and low-side pressure too high
   B. No pressure on either gauge
   C. Pressure readings that are too low for both gauges
   D. Pressure readings that are too high for both gauges

TASKS B.2.1, B.2.6

   Answer A is correct. A compressor with worn reed valves would be unable to create very much pressure difference between high- and low-side readings.

   Answer B is incorrect. No pressure would most likely mean the system is empty.

   Answer C is incorrect. Both gauges reading too low would likely indicate an undercharge or a restriction.

   Answer D is incorrect. Both gauges reading too high could indicate an overcharge, high operating temperatures, air or moisture in the system, or a TXV valve stuck open.

16. Technician A says that when installing a new or remanufactured compressor you should first turn it over by hand and drain any oil shipped with the compressor, and then install the correct amount of the specified oil before installing it. Technician B says that some compressors are shipped without oil. Who is correct?

TASK B.2.6

   A. A only
   B. B only
   C. Both A and B
   D. Neither A nor B

   Answer A is incorrect. Technician B is also correct.

   Answer B is incorrect. Technician A is also correct.

   Answer C is correct. Both Technicians are correct. When installing a new or rebuilt compressor, first turn it over by hand and drain any oil shipped with the compressor, and then install the correct amount of the right oil before installing it. In addition, some compressors are shipped without oil. Most manufacturers recommend draining fluid from the old compressor into a measuring cup. If two ounces or more are drained from the old compressor, then the same amount of new oil should be added to the new compressor. If less than two ounces are drained from the old compressor, then two ounces of new oil should be added to the new compressor. After adding the oil, the compressor should be turned by hand to help displace the oil to prevent damage when the compressor is first turned on.

   Answer D is incorrect. Both Technicians are correct.

TASK B.2.3

17. A knocking noise is heard in the compressor area that is audible when the compressor is engaged, but that goes away when the compressor turns off. Technician A says that loose compressor mounting bolts could be the cause. Technician B says that a discharge line rubbing a compressor mounting bracket could be the cause. Who is correct?

   A. A only
   B. B only
   C. Both A and B
   D. Neither A nor B

Answer A is incorrect. Technician B is also correct.

Answer B is incorrect. Technician A is also correct.

Answer C is correct. Both Technicians are correct. Loose compressor mounting bolts or a discharge line rubbing a bracket could both cause a knocking when the compressor is engaged.

Answer D is incorrect. Both Technicians are correct.

TASK B.3.4

18. A routine A/C maintenance service should include all of the following EXCEPT:

   A. Tightening the condenser lines.
   B. Removing debris from the condenser fins.
   C. Straightening the condenser fins.
   D. Checking the condenser mounts.

Answer A is correct. A/C fittings should not be disturbed if they are not leaking.

Answer B is incorrect. Cleaning the condenser fins is recommended for an A/C maintenance service.

Answer C is incorrect. An A/C maintenance service should include straightening bent condenser fins.

Answer D is incorrect. Checking all component mounts and insulators should be part of an A/C maintenance service.

TASK A.5

19. All of the following features could be found on a truck with an automatic climate control system EXCEPT:

   A. Cabin temperature sensor.
   B. Electronic climate control head.
   C. HVAC logic device.
   D. Blower resistor.

Answer A is incorrect. Automatic climate control systems use a cabin temperature sensor to signal the control module the temperature in the cabin.

Answer B is incorrect. Automatic climate control systems use an electronic climate control head to allow the driver or passenger to select the desired cabin temperature. The system uses the selected temperature input to signal the climate control module what temperature to achieve.

Answer C is incorrect. Automatic climate control systems use a logic device such as a computer or module to provide the control function of this advanced climate control system.

Answer D is correct. Automatic climate control systems do not use a blower resistor. These systems use a blower control module to provide a very wide range of blower speeds depending on the signal from the climate control module.

20. All of the following statements about coolant control valves are true EXCEPT:

    A. It controls the flow of coolant through the heater core.

    B. It is part of the water pump.

    C. It may be cable-operated.

    D. It may be vacuum-operated.

TASK C.10

Answer A is incorrect. The coolant control valve does control the flow of coolant through the heater core.

Answer B is correct. The coolant control valve is not part of the water pump. It is usually located in the inlet heater hose somewhere near the firewall.

Answer C is incorrect. The coolant control valve may be cable-operated.

Answer D is incorrect. The coolant control valve may be vacuum-operated.

21. Which of the following is the location for the expansion valve?

    A. Inlet line of the evaporator

    B. Outlet line of the evaporator

    C. Inlet line of the compressor

    D. Outlet line of the condenser

TASK B.3.6

Answer A is correct. The expansion valve is located at the evaporator inlet.

Answer B is incorrect. The evaporator outlet is where the accumulator dryer would be located.

Answer C is incorrect. The outlet line of the compressor is known as the discharge line.

Answer D is incorrect. This is the incorrect location. The receiver/drier is located in this location.

22. The A/C compressor high-pressure relief valve:

    A. Is calibrated by shimming it to the proper depth.

    B. Must be replaced if it ever vents refrigerant from the system.

    C. Will reset itself when A/C system pressure returns to a safe level.

    D. Is not used in R-134a systems.

TASK B.3.10

Answer A is incorrect. The relief valve cannot be calibrated.

Answer B is incorrect. The relief valve does not need to be replaced if it vents refrigerant to the atmosphere and then resets itself.

Answer C is correct. The valve will reset itself when A/C system pressure returns to a safe level. The valve is designed to release pressure at 450–550 psi (3103–3792 kPa).

Answer D is incorrect. All mobile A/C systems use a high-pressure relief valve.

TASK A.5

23. Which of the following features is LEAST LIKELY to be found on a truck with an automatic climate control?

    A. Cable-operated mode door actuator
    B. Cabin temperature sensor
    C. Electronic climate control head
    D. Electronic blend door actuator

    Answer A is correct. Automatic climate control systems do not use cable-operated mode actuators. These systems must use an electronic actuator with a feedback device that sends a signal to the climate control module revealing the position of the mode door.

    Answer B is incorrect. Automatic climate control systems use a cabin temperature sensor to signal the control module the temperature in the cabin.

    Answer C is incorrect. Automatic climate control systems use an electronic climate control head to allow the driver or passenger to select the desired cabin temperature. The system uses the selected temperature input to signal the climate control module what temperature to achieve.

    Answer D is incorrect. Automatic climate control systems use electronic blend door actuators that have a feedback device that sends a signal to the climate control module revealing the position of the blend door.

**TASKS C.6,
C.7**

24. The LEAST LIKELY cause of poor coolant circulation in a truck with a down-flow radiator is:

    A. A defective thermostat.
    B. An eroded water pump impeller.
    C. A collapsed upper radiator hose.
    D. A collapsed lower radiator hose.

    Answer A is incorrect. A defective thermostat can cause poor coolant circulation.

    Answer B is incorrect. An eroded water pump impeller will cause poor coolant circulation.

    Answer C is correct. The upper hose is usually under pressure and is unlikely to collapse.

    Answer D is incorrect. The lower hose is on the suction side of the water pump and could collapse, even though it usually contains a spring to keep this from happening.

TASK C.8

25. All of these statements about cooling system service are true EXCEPT:

    A. When the cooling system pressure is increased, the boiling point is decreased.
    B. If more antifreeze is added to the coolant mix, the boiling point is increased.
    C. A good quality ethylene glycol antifreeze contains a corrosion inhibitor.
    D. Coolant solutions must be recovered, recycled, or handled as hazardous material.

    Answer A is correct. When you increase the cooling system pressure, the boiling point is increased, not decreased.

    Answer B is incorrect. When more antifreeze is added to the coolant mix, the boiling point is increased.

    Answer C is incorrect. High quality ethylene glycol antifreeze contains a corrosion inhibitor.

    Answer D is incorrect. Coolant solutions must be recovered, recycled, or handled as hazardous material.

26. Technician A says that a cracked fan blade should be welded. Technician B says that a cracked fan blade can be repaired with epoxy. Who is correct?

TASK C.9

    A. A only
    B. B only
    C. Both A and B
    D. Neither A nor B

    Answer A is incorrect. A cracked fan blade should be replaced, not welded. There is too much danger that the weld could break loose and damage the radiator or even injure a person near the truck.

    Answer B is incorrect. A technician should never attempt to repair a cracked fan blade.

    Answer C is incorrect. Neither Technician is correct.

    Answer D is correct. Neither Technician is correct. Cracked fan blades are never to be welded or repaired with epoxy. Always replace a cracked blade with a new one.

27. Technician A says that you should replace the receiver/drier if the sight glass appears cloudy because it indicates a ruptured desiccant pack. Technician B says that you should only replace the receiver/drier if it has a leak. Who is correct?

TASK B.3.5

    A. A only
    B. B only
    C. Both A and B
    D. Neither A nor B

    Answer A is correct. Only Technician A is correct. A cloudy sight glass indicates that the desiccant pack in the receiver/drier has broken.

    Answer B is incorrect. The receiver/drier must be replaced if the desiccant pack breaks down or if the system has been open for an extended period of time.

    Answer C is incorrect. Only Technician A is correct.

    Answer D is incorrect. Technician A is correct.

28. Which of the following statements is the LEAST LIKELY result of a touch test performed on a normal A/C system?

TASK A.3

    A. The compressor discharge line is hot to the touch.
    B. The line exiting the orifice tube is cold with a frost ring around it.
    C. The suction line is cold with condensation droplets on it.
    D. The line exiting the condenser is not as hot as the line entering the condenser.

    Answer A is incorrect. The compressor discharge line is normally hot to the touch. Higher ambient temperature will result in hotter discharge line temperatures.

    Answer B is correct. The line exiting the orifice tube should not be cold enough to have a frost ring around it. Frost at this location would indicate a restricted orifice tube.

    Answer C is incorrect. The suction line should be cold and will have condensation droplets if there is any humidity present in the outside air.

    Answer D is incorrect. The condenser outlet line should be about 20° F cooler than the inlet line.

TASK D.3

29. What does the pressure cycling switch located on the accumulator sense?

    A. Outside temperature
    B. Accumulator pressure
    C. Accumulator temperature
    D. Engine compartment temperature

Answer A is incorrect. The cycling switch does not sense outside temperature.

Answer B is correct. The cycling switch is mounted in the accumulator where it senses pressure. The cycling switch usually opens at about 20–25 psi (137–172 kPa) and closes at about 40–45 psi (276–310 kPa).

Answer C is incorrect. The cycling switch senses accumulator pressure, not temperature.

Answer D is incorrect. The cycling switch does not sense engine ambient temperature.

TASKS D.4, D.6

30. A coolant temperature sensor is classified as negative temperature coefficient (NTC). Technician A says this is a thermistor in which internal resistance decreases in proportion to temperature rise. Technician B says that in most truck engine cooling systems, a thermistor is supplied with battery voltage (V-Bat) and returns a portion of it as a signal. Who is correct?

    A. A only
    B. B only
    C. Both A and B
    D. Neither A nor B

Answer A is correct. Only Technician A is correct. An NTC thermistor is commonly used as a coolant temperature sensor in truck engines. NTC stands for negative temperature coefficient. As the temperature rises, the resistance decreases.

Answer B is incorrect. Truck coolant sensors are usually supplied with V-Ref (± 5 VDC) and not V-Bat.

Answer C is incorrect. Only Technician A is correct.

Answer D is incorrect. Technician A is correct.

TASK D.6

31. A truck is being diagnosed for a problem with the blend door actuator motor. The motor runs when the temperature setting is changed, but the blend door does not move. Which of the following conditions would be the most likely cause of this problem?

    A. Defective control module
    B. Defective actuator feedback device
    C. Defective drive gear in the actuator
    D. Improperly adjusted mode door linkage

Answer A is incorrect. The running motor indicates that the module is functioning.

Answer B is incorrect. A defective feedback device will not prevent the door from moving when the motor runs. If the potentiometer is bad, the motor will run, but the blend door will not be in the proper position.

Answer C is correct. A defective drive gear in the actuator will cause this condition. This problem is usually accompanied by a clicking noise from the actuator in the dash area. The actuator must be replaced to correct this problem.

Answer D is incorrect. An improperly adjusted mode door linkage would not cause a temperature performance problem.

32. All of the statements about replacing an HVAC control panel are true EXCEPT:

    A. The negative battery cable should be removed before servicing the control panel.

    B. The refrigerant must be recovered before removing the control panel.

    C. If the truck contains a supplemental restraint system (SRS), a technician must wait the specified period after removing the negative battery cable.

    D. Self-diagnostic tests may indicate a defective control panel in an automatic temperature control (ATC) system.

TASK D.7

Answer A is incorrect. It is advisable to remove the negative battery cable before control panel service.

Answer B is correct. A technician does not have to recover the refrigerant before removing the control panel.

Answer C is incorrect. If the truck contains a supplemental restraint system, a technician must wait the specified period after removing the negative battery cable.

Answer D is incorrect. Self-diagnostic tests may indicate a defective control panel in an ATC system.

33. Technician A says a container of PAG refrigerant oil must be kept closed when not in use to prevent the oil from absorbing moisture. Technician B says that oil must be added to all new components during installation. Who is correct?

    A. A only

    B. B only

    C. Both A and B

    D. Neither A nor B

TASK B.1.2

Answer A is incorrect. Technician B is also correct.

Answer B is incorrect. Technician A is also correct.

Answer C is correct. Both Technicians are correct. PAG oil is very hydroscopic, so it should be kept closed tightly when not in use. It is a good practice to purchase refrigerant oil in small containers so that less oil is wasted due to moisture absorption. Refrigerant oil should be added to each component as it is replaced in order to spread out the oil to various parts of the system.

Answer D is incorrect. Both Technicians are correct.

34. All of the following statements about air-controlled HVAC systems in heavy-duty trucks are true EXCEPT:

    A. Air cylinders are used to open shutters.

    B. Coolant control valves can be controlled using chassis air.

    C. Air cylinders are used to control mode and blend air doors.

    D. Air leaks may cause mode doors to move slowly or to be totally inoperative.

TASK D.8

Answer A is correct. Air cylinders are used to close shutters; they are opened by spring force.

Answer B is incorrect. Coolant control valves can be operated by chassis air.

Answer C is incorrect. In air-controlled HVAC systems, mode and blend air cylinders typically operate air doors.

Answer D is incorrect. Air leaks can cause mode and blend air doors to react sluggishly or to be inoperative.

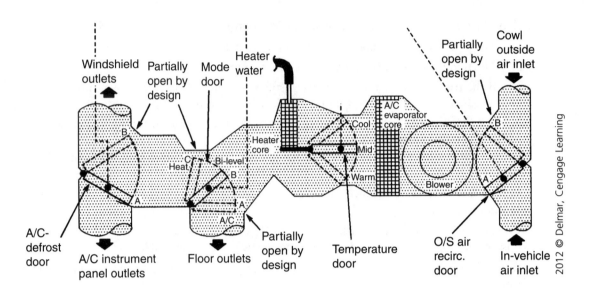

TASKS D.6, D.7

35. Referring to the figure above, the outside air recirculation door is stuck in position A. Technician A says under this condition outside air is drawn into the HVAC inlet. Technician B says under this condition in-vehicle air is recirculated. Who is correct?

A.   A only

B.   B only

C.   Both A and B

D.   Neither A nor B

Answer A is correct. Only Technician A is correct. When the recirculation door is in position A, outside air is drawn into the HVAC case. A noticeable change in airflow should occur when the fresh/recirculate control is changed.

Answer B is incorrect. In position A, in vehicle air is blocked.

Answer C is incorrect. Only Technician A is correct.

Answer D is incorrect. Technician A is correct.

36. The best instrument to use when troubleshooting an A/C electronic circuit with solid-state components is:

TASK D.10

   A.   A digital multi-meter (DMM).

   B.   A self-powered test lamp.

   C.   An analog volt/ohmmeter (VOM).

   D.   A 12 V test lamp.

   Answer A is correct. A DMM is the most precise tool for electrical/electronic components or systems, and because of its 10 megohm impedance it will not harm solid-state components.

   Answer B is incorrect. A self-powered test lamp could damage solid-state components. The self-powered test light, also called a continuity tester, is only used to check switches and wiring for continuity.

   Answer C is incorrect. An analog VOM can damage solid-state components because it draws too much current.

   Answer D is incorrect. A 12 V test lamp can damage solid-state components because it draws too much current.

37. A refrigerant identifier is connected to a truck and gives readings of 100 percent R-134a and 0.0 percent R-12. Technician A says that this system can be safely recovered into the R-134a recovery machine. Technician B says that refrigerant identifiers should be used on every vehicle prior to connecting any A/C equipment. Who is correct?

   TASK B.1.5

   A.   A only

   B.   B only

   C.   Both A and B

   D.   Neither A nor B

   Answer A is incorrect. Technician B is also correct.

   Answer B is incorrect. Technician A is also correct.

   Answer C is correct. Both Technicians are correct. An R-134a system that has no other substances in it can safely be recovered into the recovery machine. It is a good idea to use the identifier on every vehicle to protect the equipment from drawing a refrigerant that is not pure.

   Answer D is incorrect. Both Technicians are correct.

Pressure-Temperature Relationship				
Temperature °F (°C)	R-12 PSIG	(bar/kg/cm2)	R-134A PSIG	(bar/kg/cm2)
–15 (–26.1)	2.5	(.17/.18)	0	(0)
–10 (–23.3)	4.5	(.31/.32)	2.0	(.14/.14)
–5 (–20.5)	6.7	(.46/.03)	4.1	(.28/.29)
0 (–17.8)	9.2	(.63/.65)	6.5	(.45/.46)
5 (–15.0)	11.8	(.81/.83)	9.1	(.63/.64)
10 (–12.2)	14.7	(1.0/1.0)	12.0	(.89/.84)
15 (–9.4)	17.7	(1.2/1.2)	15.1	(1.0/1.2)
20 (–6.7)	21.1	(1.5/1.5)	18.4	(1.3/1.3)
25 (–3.9)	24.6	(1.7/1.7)	22.1	(1.5/1.6)
30 (–1.1)	28.5	(2.0./2.0)	26.1	(1.8/1.8)
35 (1.7)	32.6	(2.2/2.3)	30.4	(2.1/2.1)
40 (4.4)	37.0	(2.6/2.6)	35.0	(2.4/2.5)
45 (7.2)	41.7	(2.9/3.0)	40.0	(2.8/2.8)
50 (10.0)	46.7	(3.2/3.3)	45.4	(3.1/3.2)
55 (12.8)	52.1	(3.6/3.7)	51.2	(3.5/3.6)
60 (15.6)	57.8	(4.0/4.1)	57.4	(4.0/4.0)
65 (18.3)	63.8	(4.4/4.5)	64.0	(4.4/4.5)
70 (21.1)	70.2	(4.8/5.0)	71.1	(5.0/5.0)
75 (23.9)	77.0	(5.3/5.4)	78.6	(5.4/5.5)
80 (26.7)	84.2	(5.8/6.0)	86.7	(6.0/6.1)
85 (29.4)	91.7	(6.3/6.4)	95.2	(6.6/6.7)
90 (32.2)	99.7	6.9/7.0)	104.3	(7.2/7.3)
95 (35.0)	108.2	(7.5/7.6)	113.9	(7.9/8.0)
100 (37.8)	117.0	(8.1/8.2	124.1	(8.6/8.7)
105 (40.6)	126.4	(8.7/8.9)	134.9	(9.3/9.5)
110 (43.3)	136.2	(9.4/9.6)	146.3	(10.1/10.3)
115 (46.1)	146.5	(10.1/10.3)	158.4	(11.0/11.1)
120 (48.9)	157.3	(11.0/11.1)	171.1	(11.8/12.0)

TASK B.1.11

38. Referring to the table above, a storage container or A/C system containing R-134a (at rest) and subject to an ambient temperature of 70° F (21° C) will have an internal gauge pressure of approximately:

A. 220 psi (1517 kPa).

B. 125 psi (862 kPa).

C. 71 psi (490 kPa).

D. 30 psi (207 kPa).

Answer A is incorrect. If a cylinder had 220 psi of pressure at 70° F, it would definitely have some other chemical mixed with it. A technician would need to use an identifier to find out what else is mixed with the refrigerant.

Answer B is incorrect. A pressure reading of 125 psi in a cylinder of refrigerant at 70° F would indicate a high percentage of air. The refrigerant would need to be recovered and recycled to remove the air.

Answer C is correct. The relationship of pressure to temperature at a constant volume is direct, so at 70° F the pressure is about 71 psi. This is known as Charles' Gas Law.

Answer D is incorrect. The relationship of pressure to temperature at a constant volume is direct, so at 70° F the pressure is about 71 psi, not 30 psi.

2012 © Delmar, Cengage Learning

39. Which of the following repair processes is the best description of the term recovery?

TASK B.1.5

 A. The process of pulling the A/C system into a vacuum in order to remove the air and moisture from the system.

 B. The process of removing and weighing the refrigerant from an A/C system.

 C. The process of filtering the refrigerant to remove impurities from it.

 D. The process of adding refrigerant to an A/C system after repair work has been performed.

Answer A is incorrect. Evacuation is the process of pulling the A/C system into a vacuum in order to remove the air and moisture from the system.

Answer B is correct. Recovery is the process of removing and weighing the refrigerant from an A/C system.

Answer C is incorrect. Recycling is the process of filtering the refrigerant to remove impurities from it.

Answer D is incorrect. Recharging is the process of adding refrigerant to an A/C system after repair work has been performed.

40. Which of the following problems would most likely cause a loud hissing sound accompanied by a momentary release of refrigerant vapor under the hood?

TASKS A.2, B.3.3

 A. Blocked thermal expansion valve

 B. Condenser coated with mud and debris

 C. Restricted evaporator core

 D. Faulty compressor

Answer A is incorrect. A blocked thermal expansion valve would not cause a loud hissing sound. This problem would cause decreased operating pressures in both sides of the A/C system, as well as poor A/C performance.

Answer B is correct. A condenser that is coated with mud and debris would cause the high side pressures to be very high, which could cause the pressure relief valve to exhaust the pressure. This pressure release would cause a loud hissing sound accompanied by a momentary release of refrigerant vapor under the hood.

Answer C is incorrect. A restricted evaporator core would not cause a loud hissing sound. This problem would cause decreased operating pressures in both sides of the A/C system as well as poor A/C performance.

Answer D is incorrect. A faulty compressor would not cause a loud hissing sound. This problem would cause both sides of the A/C system to have equal pressures.

PREPARATION EXAM ANSWER SHEET FORMS

ANSWER SHEET

1. _____	21. _____
2. _____	22. _____
3. _____	23. _____
4. _____	24. _____
5. _____	25. _____
6. _____	26. _____
7. _____	27. _____
8. _____	28. _____
9. _____	29. _____
10. _____	30. _____
11. _____	31. _____
12. _____	32. _____
13. _____	33. _____
14. _____	34. _____
15. _____	35. _____
16. _____	36. _____
17. _____	37. _____
18. _____	38. _____
19. _____	39. _____
20. _____	40. _____

ANSWER SHEET

1. _____

2. _____

3. _____

4. _____

5. _____

6. _____

7. _____

8. _____

9. _____

10. _____

11. _____

12. _____

13. _____

14. _____

15. _____

16. _____

17. _____

18. _____

19. _____

20. _____

21. _____

22. _____

23. _____

24. _____

25. _____

26. _____

27. _____

28. _____

29. _____

30. _____

31. _____

32. _____

33. _____

34. _____

35. _____

36. _____

37. _____

38. _____

39. _____

40. _____

ANSWER SHEET

1. _____
2. _____
3. _____
4. _____
5. _____
6. _____
7. _____
8. _____
9. _____
10. _____
11. _____
12. _____
13. _____
14. _____
15. _____
16. _____
17. _____
18. _____
19. _____
20. _____

21. _____
22. _____
23. _____
24. _____
25. _____
26. _____
27. _____
28. _____
29. _____
30. _____
31. _____
32. _____
33. _____
34. _____
35. _____
36. _____
37. _____
38. _____
39. _____
40. _____

ANSWER SHEET

1. _____

2. _____

3. _____

4. _____

5. _____

6. _____

7. _____

8. _____

9. _____

10. _____

11. _____

12. _____

13. _____

14. _____

15. _____

16. _____

17. _____

18. _____

19. _____

20. _____

21. _____

22. _____

23. _____

24. _____

25. _____

26. _____

27. _____

28. _____

29. _____

30. _____

31. _____

32. _____

33. _____

34. _____

35. _____

36. _____

37. _____

38. _____

39. _____

40. _____

ANSWER SHEET

1. _____   21. _____
2. _____   22. _____
3. _____   23. _____
4. _____   24. _____
5. _____   25. _____
6. _____   26. _____
7. _____   27. _____
8. _____   28. _____
9. _____   29. _____
10. _____   30. _____
11. _____   31. _____
12. _____   32. _____
13. _____   33. _____
14. _____   34. _____
15. _____   35. _____
16. _____   36. _____
17. _____   37. _____
18. _____   38. _____
19. _____   39. _____
20. _____   40. _____

ANSWER SHEET

1. _____
2. _____
3. _____
4. _____
5. _____
6. _____
7. _____
8. _____
9. _____
10. _____
11. _____
12. _____
13. _____
14. _____
15. _____
16. _____
17. _____
18. _____
19. _____
20. _____

21. _____
22. _____
23. _____
24. _____
25. _____
26. _____
27. _____
28. _____
29. _____
30. _____
31. _____
32. _____
33. _____
34. _____
35. _____
36. _____
37. _____
38. _____
39. _____
40. _____

Glossary

Absolute Pressure The zero point from which pressure is measured.

Access Valve A term used for service port and service valve.

Accumulator A tank located between the evaporator and compressor to receive the refrigerant that leaves the evaporator, so constructed as to ensure that no liquid refrigerant enters the compressor.

Actuator A device that transfers a vacuum or electric signal to a mechanical motion, typically performs an on/off or open/close function. Used in HVAC systems to control the blend air door and the mode doors.

A/C Pressure Transducer Also called the pressure sensor. A device that senses pressure in the high side by varying voltage levels.

Adapter A device or fitting that permits different size parts or components to be fastened or connected to each other.

Additive An ingredient intended to improve a certain characteristic of a material or fluid.

After-Cooler A heat exchanger often cooled by engine coolant.

Air Conditioning (A/C) The control of air movement, humidity, and temperature by thermodynamics.

Air Drier A device that removes moisture.

Air Gap The space between two components, such as between the rotor and armature of a clutch.

Ambient Sensor A thermistor used in automatic temperature control units to sense ambient temperature, which is the temperature of the surrounding or prevailing air, as in the service area where testing is taking place.

Ampere The unit for measuring electrical current, often referred to simply as amp.

Analog Volt/Ohmmeter (AVOM) A test meter used for checking voltage and resistance. Analog meters should not be used on solid-state circuits.

Antifreeze A compound, such as ethylene or propylene glycol, added to water to lower its freezing point.

Armature The part of the clutch that mounts onto the crankshaft and engages with the rotor when energized; the rotating component of a (1) starter or other motor, (2) generator, or (3) compressor clutch.

ASE Automotive Service Excellence, a trademark of the National Institute for Automotive Service Excellence (formerly known by the acronym NIASE).

Atmospheric Pressure Air pressure at a given altitude. At sea level, atmospheric pressure is 14.696 psi (101.329 kPa absolute).

Automatic Temperature Control (ATC) A type of heating and air conditioning system that allows the driver to select a desired temperature on the control head. The A/C computer calculates the necessary output to adjust the cabin temperature.

Barb Fitting A fitting that slips inside a hose and is held in place with a gear-type clamp. Ridges (barbs) on the fitting prevent the hose from slipping off.

BCM An abbreviation for body control module.

Binary Pressure Switch A switch that prevents compressor operation if the refrigerant change has been lost or ambient temperature is too cold; protects the system from excessive pressure. Also called the dual pressure switch.

Blend Air Door The main device that controls the air temperature by routing the air through or around the heater core. Also called the temperature blend door.

Blend Air System The system by which air flowing through the HVAC duct is channeled by the blend air door to flow through or around the heater core, depending on the temperature lever setting.

Blend Door Actuator An electric device that controls the temperature blend door.

Blower Fan A fan that blows air through a ventilation, heater, or air conditioning system.

Blower Relay An electrical device used to control the function or speed of a blower motor.

Blower Resistor Block An assembly that contains several resistors in a series and is used to step down the voltage to the blower motor, thereby providing multiple blower speeds.

Boiling Point The temperature at which a liquid changes to a vapor, as opposed to the condensation point, the temperature at which a vapor changes to a liquid.

Break a Vacuum The next step after evacuating a system. The vacuum should be broken with refrigerant or other suitable dry gas, not ambient air or oxygen.

British Thermal Unit (Btu) A measure of heat quantity; the amount of heat required to raise 1 pound of water by one degree F.

By-Pass An alternate passage that may be used instead of the main passage.

By-Pass Hose A hose that is generally small and is used as an alternate passage to by-pass a component or device.

Can Tap A device used to pierce, dispense, and seal small cans of refrigerant.

Can Tap Valve A valve found on a can tap that is used to control the flow of refrigerant.

Cap (1) A protective cover. (2) An abbreviation for capillary (tube) or capacitor.

Cap Tube A tube with a calibrated inside diameter and length used to control the flow of refrigerant between the remote bulb and the expansion valve.

CCOT Abbreviation for cycling clutch orifice tube.

Celsius (C) A metric temperature scale using zero as the freezing point of water. The boiling point of water is 100° C (212° F).

Certified Having a certificate awarded or issued to those that have demonstrated appropriate competence through testing and/or practical experience.

CFC-12 A term used for Refrigerant-12.

Charge A specific amount of refrigerant or oil by volume or weight.

Check Valve A device that prevents refrigerant from flowing in the opposite direction when the unit is shut; it allows fluid to flow in one direction only.

Clean Air Act (CCA) A Title IV amendment signed into law in 1990 that established national policy relative to the reduction and elimination of ozone-depleting substances.

Chlorofluorocarbon (CFC) A compound used in the production of some refrigerants that damage the ozone layer.

Circuit The complete path of an electrical current, including the generating device. When the path is unbroken, the circuit is closed and current flows. When the circuit continuity is broken, the circuit is open and current flow stops.

Clutch An electro-mechanical device mounted on the air conditioning compressor used to start and stop compressor action, thereby controlling refrigerant circulating through the system.

Clutch Coil The electrical part of a clutch assembly. When electrical power is applied to the clutch coil, the clutch is engaged to start and stop compressor action.

Compound Gauge A gauge that registers both pressure and vacuum (above and below atmospheric pressure); used on the low side of the systems.

Compressor Component of an air conditioning system that compresses low-pressure refrigerant vapor and pumps it through the refrigeration circuit.

Compressor-Shaft Seal An assembly consisting of springs, snap rings, o-rings, shaft seal, seal sets, and gasket, mounted on the compressor crankshaft to permit the shaft to be turned without a loss of refrigerant or oil.

Condenser Component in an air conditioning system used to cool refrigerant to below its boiling point, changing it from a vapor to a liquid, through the process of condensation.

Contaminated A term used when referring to a refrigerant cylinder or a system that is known to contain foreign substances, such as other incompatible or hazardous refrigerants.

Coolant Liquid that circulates in an engine cooling system, usually a solution of antifreeze and water.

Coolant Flow Control System The system by which temperature control is achieved using a coolant control valve (hot-water valve) to limit the amount of coolant allowed to flow through the heater core.

Coolant Heater A component used to aid engine starting and reduce the wear caused by cold starting.

Coolant Hydrometer A tester designed to measure coolant-specific gravity and to determine the amount of antifreeze protection.

Cooling System Engine system for circulating coolant.

Cracked Position A mid-seated or open position.

Cut-Out Switch The low-pressure cut-out switch interrupts compressor operation if system pressure drops to the point that a loss of refrigerant charge to the compressor occurs. The high-pressure cut-out switch interrupts compressor operation under situations of extremely high system pressure. Also called the pressure cycling switch or cycling clutch pressure switch.

Cycle Clutch Time (Total) Time elapsed from the moment the clutch engages until it disengages, then reengages. Total time is equal to on-time plus off-time for one cycle. Cycling is the repeated on-off action of the air conditioner.

Cycling Clutch Pressure Switch A pressure-actuated electrical switch used to cycle the compressor at a predetermined pressure.

Cycling Clutch System An air conditioning system in which the air temperature is controlled by starting and stopping the compressor with a thermostat or pressure control.

Data Link Connector (DLC) Contact point through which computers communicate with other electronic devices, such as control panels, modules, sensors, or other computers.

Department of Transportation (DOT) A cabinet-level Federal government agency that establishes standards for vehicles, road systems, and the regulation and control of the shipment of all hazardous materials.

Depressing Pin A pin located in the end of a service hose to press (open) a Schrader-type valve.

Diagnostic Flow Chart A model-specific chart found in the service manual that provides a systematic approach to component troubleshooting and repair.

Diagnostic Trouble Code (DTC) Diagnostic trouble code in some electronic systems.

Digital Diagnostic Reader (DDR) An electronic service tool usually handheld or PC-based.

Digital Multi-Meter (DMM) A device used in testing voltage, amperage, resistance, and continuity of electrical circuits and components. This term is used more commonly, albeit interchangeably, with digital volt ohmmeter (DVOM).

Digital Volt Ohmmeter (DVOM) A testing device, also known as digital multi-meter (DMM).

Dual Pressure Switch Also called the binary pressure switch, it opens at high pressure.

Duct A tube or passage used to provide a means to transfer air or liquid from one point or place to another.

ECM Abbreviation for electronic control module.

ECT Abbreviation for engine coolant temperature.

EG Abbreviation for ethylene glycol, a commonly used anti-freeze agent.

ELC Abbreviation for extended life coolant, premixed coolant solution with high resistance to breakdown.

Electronic Control Module (ECM) A controller module that contains the CPU and output generators. Grounds the low- and high-speed fan relays in response to engine coolant temperature and compressor head temperature. Sometimes referred to as electronic control unit (ECU).

Environmental Protection Agency (EPA) An agency of the U.S. government that is charged with responsibility for protecting the environment and enforcing the Clean Air Act (CAA) of 1990.

Evacuate To create a vacuum within a system to remove all traces of air and moisture.

Evaporator Component in an air conditioning system used to remove heat from air forced through it.

Evaporator Case The housing containing the evaporator core, the heater core, the evaporator core drain, and the blend air and mode control doors.

Evaporator Temperature Sensor A device that senses the temperature in the evaporator. Also called the fin temp sensor.

Evaporator Temperature Switch A thermostatic switch that opens in response to cold temperatures in the evaporator.

Fan Relay A relay for the cooling and/or auxiliary fan motors.

Fault Code A trouble code that is written to computer memory. A fault code can be read by a digital diagnostic reader.

Fill Neck The part of the radiator on which the pressure cap is attached. Most radiators, however, are filled via the recovery tank.

Filter A device used with the dryer or as a separate unit to remove foreign material from the refrigerant.

Filter Drier A device that has a filter to remove foreign material from the refrigerant and a desiccant to remove moisture from the refrigerant.

Fin Temp Sensor A device that senses the temperature in the evaporator. Also called the evaporator temperature sensor.

Flammable A term used to describe any material that will catch fire or explode.

Flow Tester A device that detects the pressure of sealer that could be mixed with refrigerant.

Foot-Pound English unit of measurement for torque.

FOT Abbreviation for fixed orifice tube.

Fresh/Recirculate Air Door The door in the air duct box that allows the air to be pulled from inside the cabin in recirculate position (RECIR) or from outside the cab in fresh position.

Front Seat Closing off the line, thus leaving the compressor open to the service port fitting. This allows service to the compressor without purging the entire system. Never operate the system with the valves front seated.

Fusible Link A term used for fuse link.

Fuse Link The short length of smaller-gauge wire installed in a conductor, usually close to the power source.

Gasket A thin layer of material or composition that is placed between two machined surfaces to provide a leakproof seal between them.

Gauge A tool of a known calibration used to measure components. For example, a feeler gauge is used to measure the air gap between a clutch rotor and armature.

Graduated Container A measure such as a beaker or measuring cup that has a graduated scale for the measure of a liquid.

Ground The negatively charged side of an electrical circuit. A ground can be a wire, the negatively charged of the battery, or the vehicle chassis.

Grounded Circuit A shorted circuit that causes a current to return to the battery before it has reached its intended destination.

Hazardous Materials Any substance that is flammable, explosive, or known to produce adverse health effects in people or the environment.

HCFC Abbreviation for hydrochlorofluorocarbon refrigerant.

Header Tank The top and bottom tanks (downflow) or side tanks (crossflow) of a radiator. The tanks in which coolant is accumulated or received.

Heat Exchanger A device, such as a radiator or condenser, in which heat is transferred from one medium to another on the principle that heat moves to an object with less heat.

Heater Control Valve A valve that controls the flow of coolant into the heater core from the engine.

Heater Core A radiator-like heat exchanger located in the case/duct system through which coolant flows to provide heat to the vehicle interior.

Heavy-Duty Truck A truck that has a GVW of 26,001 pounds or more.

HI The designation for high as in blower speed or system mode.

High-Pressure Relief Valve A device that vents refrigerant from the system in the event that high-side pressure exceeds safe levels.

High-Pressure Switch The pressure cut-off switch that opens under high A/C system pressure.

High-Side Gauge The correct side gauge on the manifold used to read refrigerant pressure in the high side of the system.

High-Side Hand Valve The high-side valve on the manifold set used to control flow between the high side and service ports.

High-Side Service Valve A device located on the discharge side of the compressor; this valve permits the service technician to check the high-side pressures and perform other necessary operations.

Hot Knife A knife-like tool that has a heated blade used for separating objects, such as evaporator cases.

Hydrometer A tester designed to measure the specific gravity of a liquid.

IAT Abbreviation for intake air temperature.

Idler A pulley device that keeps the belt whip out of the drive belt of an automotive air conditioner. The idler is used as a means of tightening the belt.

In-Car Temperature Sensor A thermistor used in automatic temperature control units for sensing the in-car temperature. Also see Thermistor.

In-Line Fuse A fuse in series with the circuit usually enclosed in a small plastic fuse holder. Used as a protection device for a portion of a circuit.

Integrated Circuit A solid-state component containing diodes, transistors, resistors, capacitors, and other electronic components mounted on a single piece of material.

Jumper Wire A wire used to by-pass temporarily a circuit or components for electrical testing. A jumper wire consists of a length of wire with an alligator clip at each end.

Latent Heat The amount of heat transfer required to cause a substance to change state (e.g., from a vapor to a liquid).

Mode Door The door in the air duct box that routes the air to either the floor, the vent, or the defrost position (DEF).

National Automotive Technicians Education Foundation (NATEF) A nonprofit organization founded to standardize programs for certifying secondary and post-secondary automotive and heavy-duty truck training programs. (See: www.natef.org)

National Institute for Automotive Service Excellence (ASE) A nonprofit organization that has an established certification program for automotive, heavy-duty truck, auto body repair, engine machine shop technicians, and parts specialists. (See: www.ase.com)

NTC Abbreviation for negative temperature coefficient.

OEM Abbreviation for original equipment manufacturer.

Open Circuit An electrical circuit whose path has been interrupted or broken either accidentally (a broken wire) or intentionally (a switch turned off).

O-Ring A synthetic rubber or plastic gasket with a round- or square-shaped cross section.

Orifice Tube A device that allows low-pressure liquid to be metered into the evaporator. Orifice is the term used for a calibrated opening in a tube or pipe to regulate the flow of a fluid or liquid.

Output Driver Computer output switches in an ECM or ECU. Output drivers are located in the ECU along with the input conditioners, microprocessor, and memory.

Overcharge A condition indicated when too much refrigerant or refrigeration oil is added to the system.

Overload A condition in excess of the design criteria. An overload will generally cause the protective device such as a fuse or pressure relief to open.

PAG Acronym for polyalkylene glycol, a synthetic oil commonly used in R-134a systems.

Performance Test Readings of the temperature and pressure under controlled conditions to determine if an air conditioning system is operating at full efficiency.

PG Acronym for propylene glycol, a commonly used antifreeze agent; not for mixture with EG.

Piercing Pin The part of a saddle valve that is used to pierce a hole in the tubing.

Pin-Type Connector A single or multiple electrical connector that is round- or pin-shaped and fits inside a matching connector.

Polyester (ESTER) A synthetic oil-like lubricant that is occasionally recommended for use in an HFC-134a system. This lubricant is compatible with both HFC-134a and CFC-12.

Positive Pressure Any pressure above atmospheric pressure.

Potentiometer A position sensor used for door position feedback.

Pound of Refrigerant A term often used by technicians when referring to a small can of refrigerant, although it actually contains less than 16 ounces, which is usually the definition of "pound."

Power Module Controls the operation of the blower motor in an automatic temperature control system.

Pressure The amount of force applied to a definite area measured in pounds per square inch (psi) English or kilopascals (kPa) metric.

Pressure Cycling Switch Also called the cut-out switch.

Pressure Differential The difference in pressure between any two points of a system or a component.

Pressure Relief Valve A valve located on an air conditioning compressor or pressure vessel that opens if system pressure is exceeded.

Pressure Sensor Also called the A/C pressure transducer. A device that senses pressure in the high side by varying voltage levels.

Printed Circuit Board Electronic circuit board made of thin, nonconductive material onto which conductive metal, such as copper, has been deposited. The metal is then etched by acid, leaving metal lines that form conductive paths for the various circuits on the board. A printed circuit board can hold many complex circuits in a small area.

Programmable Read Only Memory (PROM) A computer memory component that contains program information specific to calibration.

psi Abbreviation for pounds per square inch and measurable with a psi gauge (PSIG).

Pulse Width (PW) The measurement of an electric circuit on time. Injector on time is typically referred to as pulse width.

Purge To remove moisture and/or air from a system or a component by flushing with a dry gas such as nitrogen (N) to remove all refrigerant from the system.

Purity Test A static test that may be performed to compare the suspect refrigerant pressure to an appropriate temperature chart to determine its purity.

Radiation The transfer of heat without heating the medium through which it is transmitted.

Ram Air Air that is forced through the radiator and condenser coils by the movement of the vehicle or the action of the fan.

Random Access Memory (RAM) The memory used during computer operation to store temporary information. The microcomputer can write, read, and erase information from RAM in any order, which is why it is called random.

RCRA Resource Conservation and Recovery Act.

Read Only Memory (ROM) A type of memory used in microcomputers to store information permanently.

Receiver/Drier A tank-like vessel having a desiccant and used for the storage of refrigerant. It is located near the condenser outlet that filters refrigerant and is also called the filter or dehydrator.

Reclamation The process used to restore used refrigerants to new product quality. It may include processes available at reprocessing facilities such as distilling and chemical analysis.

Recovery The process of removing refrigerant from a system and storing it in an external container for further processing or reuse.

Recovery System A term often used to refer to the circuit inside the recovery unit used to recycle and/or transfer refrigerant from the air conditioning system to the recovery cylinder.

Recovery Tank An auxiliary tank, usually connected to the inlet tank of a radiator, which provides additional storage space for heated coolant.

Recycling The process of cleaning and removing oil, moisture, and acidity from refrigerant, usually at a repair shop, prior to its reuse.

Reed Valves Mechanical valves that separate the low side and high side of the A/C system.

Reference Voltage The voltage supplied to a sensor by the computer, which acts as a baseline voltage. It is modified by the sensor to act as an input signal. Usually 5 VDC.

Refractometer A device that refracts light through a liquid to measure specific gravity; used to measure antifreeze protection.

Refrigerant A liquid capable of vaporizing at a low temperature.

Refrigerant-12 A refrigerant used in automotive air conditioners, as well as other air conditioning and refrigeration systems.

Refrigerant Management Center Equipment designed to recover, recycle, and recharge an air conditioning system.

Relay An electrical switch device that is activated by a low-current source and controls a high-current device. A relay consists of a control circuit and a power circuit.

Reserve Tank A storage vessel for excess fluid. See Recovery Tank, Receiver/Drier, and Accumulator.

Resistor A voltage-dropping device that is usually wire wound and provides a means of controlling fan speeds.

Respirator A mask or face shield worn in a hazardous environment to provide clean fresh air and/or oxygen.

Restrictor An insert fitting or device used to control the flow of refrigerant or refrigeration oil.

Retrofitting The name given to the procedure for converting R-12 A/C systems to be able to use R-134a refrigerant.

Right to Know Law A law passed by the federal government and administered by the Occupational Safety and Health Administration (OSHA) that requires any company that uses or produces hazardous chemicals or substances to inform its employees, customers, and vendors of any potential hazards that may exist in the workplace as a result of using the products.

Rotor The rotating or freewheeling portion of a clutch; the belt slides on the rotor.

Saddle Valve A two-part accessory valve that may be clamped around the metal part of a system hose to provide access to the air conditioning system for service.

SAE Abbreviation for Society of Automotive Engineers, a professional organization of the automotive industry and founded in 1905. The SAE is dedicated to providing technical information and standards to the automotive industry.

Schrader Valve A spring-loaded valve similar to a tire valve. The Schrader valve is located inside the service valve fitting and is used on some control devices to hold refrigerant in the system. Special adapters must be used with the gauge hose to allow access to the system.

Seal Generally refers to a compressor shaft oil seal; matching shaft-mounted seal face and front head-mounted seal seat to prevent refrigerant and/or oil from escaping. May also refer to any gasket or o-ring used between two mating surfaces for the same purpose.

Seal Seat The part of a compressor shaft seal assembly that is stationary and matches the rotating part, known as the seal face or shaft seal.

Semi-Automatic Temperature Control (SATC) A system that regulates the temperature of the output air and relies on the driver to select the desired mode and blower speed.

Sensor An electronic device used to monitor conditions for input to a computer.

Serpentine Belt A flat or V-groove belt that winds through all of the engine accessories to drive them off the crankshaft pulley; sometimes called a poly belt.

Service Port A fitting found on the service valves and some control devices; the manifold set hoses are connected to this fitting.

Solenoid An electromagnet used to perform mechanical work, made with coil windings wound around an iron tube.

Solid State Referring to electronics, consisting of semiconductor devices and other related nonmechanical components.

Spade Fuse A term used for blade fuse.

Spade-Type Connector A single or multiple electrical connector that has flat, spade-like mating provisions.

Specialty Service Shop A shop that specializes in areas such as engine rebuilding, transmission/axle overhauling, brake, air conditioning/heating repairs, and electrical/electronic work.

Specifications Design characteristics of a component or assembly noted by the manufacturer. Specifications for a vehicle include fluid capacities, weights, and other pertinent maintenance information.

Spike In our application, an electrical spike. An unwanted momentary high-energy electrical surge.

Spring Lock Fitting A special fitting using a spring to lock the mating parts together forming a leak-proof joint.

Squirrel-Cage Blower A blower wheel designed to provide a large volume of air with a minimum of noise. The blower is more compact than the fan and air can be directed more efficiently.

Stepped Resistor A resistor designed to have two or more fixed values by connecting wires to several taps.

Sun Load Heat intensity and/or light intensity produced by the sun.

Superheat Switch An electrical switch activated by an abnormal temperature-pressure condition (a superheated vapor); used for system protection.

Supplemental Coolant Additive (SCA) A corrosion inhibitor additive.

Switch A device used to control on/off and direct current flow in a circuit. A switch can be under the control of the driver or it can be self-operating.

Tank A term used for header tank and expansion tank.

Temperature Door A door within the case/duct stem to direct air through the heater and/or evaporator core.

Temperature Switch A switch actuated by a change in temperature at a predetermined point.

Thermal Cycling A process that involves letting a truck warm up and then letting it cool down again, which promotes ridding the system of air pockets.

Thermal Limiter A device located on the compressor that interrupts power to the compressor when the temperature of the compressor becomes too high.

Thermistor A device that varies its resistance as the temperature changes.

Time Guide Reference used for computing compensation payable by the truck manufacturer for repairs or service work to vehicles under warranty or for other special conditions authorized by the company.

Torque Rotary or turning force; for example, the force required to seal a connection; measured in (English) foot-pounds (ft-lb) or inch-pounds (in-lb) and in (metric) Newton-meters (N·m).

Toxicity A measure of how poisonous a substance is.

Transistor An electronic device produced by joining three sections of semiconductor materials. Used as a switching device.

Tree Diagnosis Chart A chart used to provide a sequence for what should be inspected or tested when troubleshooting a repair problem.

Triple Evacuation A process of evacuation that involves three pumpdowns and two system purges with an inert gas such as dry nitrogen (N).

TTMA Abbreviation for the Truck and Trailer Manufacturers Association.

TXV Abbreviation for thermal expansion valve.

Ultraviolet (UV) The part of the electromagnetic spectrum emitted by the sun that lies between visible violet light and X-rays.

United Nations Environmental Program (UNEP) An international initiative to mandate the eventual phase-out of CFC-based refrigerants.

Vacuum An absence of matter often used to describe any pressure below atmospheric pressure.

Vacuum Gauge A gauge used to measure below atmospheric pressure.

Vacuum Motor A device designed to provide mechanical control by the use of a vacuum.

Vacuum Pump A mechanical device used to evacuate the refrigeration system to rid it of excess moisture and air.

Validity List A list of valid bulletins supplied by the manufacturer.

V-Belt A rubber-like continuous loop placed between the engine crankshaft pulley and accessories to transfer rotary motion of the crankshaft to the accessories.

Ventilation The act of supplying fresh air to an enclosed space such as the inside of an automobile.

V-Groove Belt A term used for V-belt.

VIN Abbreviation for vehicle identification number.

Viscosity Resistance to fluid sheer. Viscosity describes oil thickness or resistance to flow.

Voltage-Generating Sensors Devices that produce their own input voltage signal.

Voltmeter A device used to measure volt(s).

VOM Abbreviation for volt ohmmeter.

Watt's Law The law of electricity used to find the power consumption in an electrical circuit expressed in watts. It states that power equals voltage multiplied by current.

Windings Coil of wire found in a relay or an electrical clutch that provides a magnetic field.

Wiring Harness A group of wires wrapped in a shroud for the distribution of power from one point to another point.

Zener Diode A variation of the diode, this device functions like a standard diode until a certain voltage is reached. When voltage level reaches this point, the Zener diode will allow current to flow in the reverse direction.